TALES FROM THE
INDIANAPOLIS 500

A COLLECTION OF THE GREATEST INDY 500 STORIES EVER TOLD

JACK ARUTE

WITH JENNA FRYER

FOREWORD BY A. J. FOYT

Sports Publishing books may be purchased in bulk at special discounts for sales promotion, corporate gifts, fund-raising, or educational purposes. Special editions can also be created to specifications. For details, contact the Special Sales Department, Sports Publishing, 307 West 36th Street, 11th Floor, New York, NY 10018 or sportspubbooks@skyhorsepublishing.com.

Visit our website at www.sportspubbooks.com

10 9 8 7 6 5 4 3 2 1

Library of Congress Cataloging-in-Publication Data available on file.

ISBN: 978-1-61321-044-4

Printed in China

For my father, Jack Arute Sr.

You are indeed the "original."

Contents

Foreword

Anyone who knows me knows that I don't have much love for the news media. I'm told they are a necessary evil, but I'm not so sure about the necessary part. Still, there are a few people out there who are fair, honest and know what they're talking about.

Jack Arute is one of 'em.

I first met Jackie at a NASCAR race back in the early 1970s when I was driving for the Wood Brothers. Back then NASCAR's top division was called Grand National. He was working as a reporter for MRN, the Motor Racing Network. To be truthful, he didn't really stand out, but with me that's probably a good thing.

It wasn't until he worked as a PR guy for Junior Johnson that I got to know him. Darrell Waltrip was Junior's driver back then, so you know I had a lot of sympathy for Jackie. Then Junior made Jackie his general manager, so he got to know what it feels like to be on the other side of the fence. I think that's one thing that makes him a good reporter. He understands the sport from the competitor's side.

When I saw him working at Indy in the early 1980s for ABC-TV, I was surprised because I had always associated him with NASCAR. He told me that he got the opportunity to work with ABC because he wasn't afraid to interview me. That didn't seem like enough of a qualification in my mind, but you never know about the news media. Or media news as I like to call them, because they're more interested in media than they are news.

I guess I can be pretty tough to deal with at times, but I just don't care to be bothered with interviews when I'm busy working. Jackie understood that. A lot of guys don't. All they think about is their job, not our job.

Another thing about Jackie is that he's pretty intense about his job. I like that. He takes it real serious, just like I take mine. He knows enough about racing that you can't blow him off with a simple answer. He doesn't just find out what, he finds out why, too. He's honest, not a phony. I like that too.

As I said, he gives me the credit for him getting his job with ABC. We were at Indy and I gave him a couple interviews, so the head honchos understood that I knew him and would work with him. His big break, as he tells it, was to get me to do The Walk with him from the garage to the grid on race morning.

He kept telling me how important it was to be on time, because it was a live show. I was scheduled to go on at exactly 10:17 a.m.

Well this was Jackie's first Indy 500 as a TV broadcaster, so that meant he was a rookie. Or, fair game!

He came by the garage about 10 minutes beforehand. I waited about five minutes and then headed to the men's room for a pit stop. I knew he'd follow because TV people are like that, but I pretended not to notice. I took my time, but I also knew when I had to be on camera.

I came out with probably less than a minute to go. Then I felt a little bad because Jackie was really nervous. I mean really nervous. Anyway, we did the interview and it came off right on time just like I knew it would.

He got the job with ABC. And he kept it even though I wasn't done playing jokes on him. One year I got hold of a water gun, it was like a bazooka. I mean it was Texas-big, and when you got hit with it, you were drenched.

Well it was a rainy day at the speedway and I was bored. Jackie was doing some live interviews right outside my garage. It was just too tempting. I let him have it right in the middle of the interview and he got soaked. I laughed my head off and so did he after he got over the shock.

And I'll tell you one thing—he is petrified, I mean really afraid of snakes. Now I'm not too crazy about them myself, but he's even more afraid of them than I am. So I always make it a point to get a piece of rope or a fake rubber snake to toss at him sometime during the month of May. It breaks up the tension in my garage to see a grown man holler and scream like a baby.

I think people will enjoy his book. He tells stories that look at Indy from the inside, the kind of stuff the average fan wouldn't know. He has a way of getting people to talk about things that they normally wouldn't talk about. His stories are short, which I also like, and it's more like he's talking to you, just like he does on TV.

The Indianapolis 500 has always been very special to me. It is the one race that makes or breaks your season. You run good at Indy, and no matter what happens the rest of the year, you feel like you've had a pretty good season. Winning there changes your life forever. I've always said, Indy is what made A.J. Foyt famous. People know me from winning there.

So when people write about Indy, I'm always a little skeptical, because I don't think they can appreciate it the way I do. But Jackie does and it comes through in his TV work and now in this book. So I hope people will read it and get to know a little more about this very special race and all that it stands for.

Of course, he isn't paying me to write this, so I guess it doesn't really matter to me if people buy the book or not. They can get it out of the library, but they should get it. They'll be happy they did.

—A. J. Foyt Jr.
Four-Time Indianapolis 500 Winning Driver

Introduction

On March 6, 2010, I received a phone call from my agent. "Jack," he said. "I don't have good news for you. NBC and IndyCar have decided not to bring you back for their telecasts this season."

I was shocked. Following the 2010 IndyCar campaign, my TV bosses told me not to worry. While a few minor changes were anticipated, my spot on the roster was secure.

"What did they say?" I asked my agent.

"They said that you were not in their plans." The fact that my agent accepted that and did not probe any deeper resulted in my seeking new representation, but to this day, that is all I know about why my almost three decade run at the Indy 500 on television ended.

That did not end my love affair with the "Greatest Spectacle in Racing." Instead, it just re-set it back to my childhood when I first discovered the 500 and fell in love with it.

Now I am just a fan. I still attend the race and celebrated with some 300,000 others its Centennial in 2011.

As I watched the pre-race ceremonies, I fantasized about what it must have been like when the first 500 Mile Sweepstakes (as it was first called) was staged.

America's romance with the automobile was in its infancy. The thought was a 500 mile race would showcase the durability of the automobile and influence sales.

What that crowd did not know was that the race winner, Ray Harroun, would change the course of driving with his innovation he employed to gain an advantage on his fellow drivers.

Those first cars required a riding mechanic. His job was to serve as a second set of eyes and ears (a co-pilot of sorts) for the driver.

As an engineer for the Marmon Auto Works located right in Indianapolis, Harroun calculated how much fuel would be needed to run the 500 miles.

He reasoned that if he could dispense with the riding mechanic, his car could increase its fuel load by the weight saved from driving solo.

Out went the riding mechanic and in went a larger gas tank. Ray also knew the need to know about the position and location of his fellow drivers, so he fashioned a mirror in a metal frame and suspended it at the top of his car's windshield.

From that day forward, every auto came equipped with a rear view mirror—an invention born from necessity.

That is the very first story to come from the Indianapolis 500, one that has survived for more than 100 years. A story that is told to visitors to the Speedway's Museum where they can still see Harroun's canary yellow car called the Marmon Wasp.

It was that story told to me as a child that first prompted me to write about my years at the Indy 500.

There are thousands of stories about Indy. It is what defines the race. They are about individuals; men and women who came to the Indianapolis Motor Speedway seeking an intangible. The stories do not define the speedway. Instead, they create the foundation for each individual's version of what Indy means.

For me, the stories that I have included were landmarks on my Indy 500 journey. Since this book was first published new stories have become part of my experience.

Stories like Helio Castroneves who has three likenesses on that iconic silver loving cup called the Borg-Warner Trophy. Stories such as Larry Foyt, son of A. J. Foyt, arguably the most famous Indy 500 driver, and his sacrifices to return his father's team to the racing pinnacle. Larry shelved his own driving career to oversee Foyt Enterprises. Along the way, he has shepherded his father's team with a keen eye to the elder Foyt's insistence upon not following the crowd. His sacrifice will not be rewarded until he returns a Foyt entry to Indy's Victory Lane.

Then, there's Dario Franchitti. A Scotsman with incredible talent who followed in the footsteps of his childhood hero, Jimmy Clark, and won the Indy 500 with the help of an expatriated Aussie named John Anderson. With weather threatening, Ando (as he was known), put Dario on a fuel conservation strategy that allowed him to stay out as others pitted. When the heavens opened up bringing out the red flag, Ando and Franchitti has won the 500. While Dario sipped the traditional quart of icy milk and soaked in his accomplishment with his wife Ashley Judd, Anderson quietly went about packing up the pit area.

Not all of Indy's stories are about the drivers or the owners. Some are simple stories about dreamers who grew up to find a place at Indy. One such story is that of Larry Curry.

He worked his way through the blue collar ranks of mechanics to eventually be the key guy at John Menard's racing outfit. He convinced Menard to hire Tony Stewart off of Indiana's short tracks.

He also went to jail.

Convicted of selling used parts from Menard's operation for his personal profit, Curry paid his debt to society and once paroled, returned to his love. Uncertain of the reception that awaited him, Curry persevered and played a key role in the development of Vision Racing and Ed Carpenter.

He has moved to Dreyer & Rinebold Racing where he oversaw their multi car team. In 2011 Curry put his 67th driver into the Indy 500 records as a starter in the famed classic. His Dreyer and Rinebold operation put four cars (the most of any team) into the Centennial edition of the 500. At the end of that race though, Curry's team watched as Dan Wheldon doused himself in milk after winning an emotional race that saw for the first time a driver crash on the last lap in the last turn and power his wrecked car down the front stretch to still finish second.

Danica Patrick's career at Indy crashed the "glass ceiling" for women. While she never won the 500 Her mark on the "Brickyard" is unmistakable. Like Tony Stewart, Danica said goodby to her Indy love at the end of 2011, to pursue a NASCAR career. Driving for, ironically, Tony Stewart and Dale Earnhardt, Jr.

Year to year and day to day, Indy's storybook adds chapters. Some are long and momentous and others are short and sweet.

But each one of them is a golden thread in the tapestry that is the Indianapolis 500.

The stories here are what Indy means to me.

—Jack Arute, November 2011

Preface

When I was first asked to write this book, I agreed only to silence folks. "Jack," they would say, "You have so many stories. You need to write a book."

I didn't think I could do it, but now that I have, I'm glad that people pestered me the way they did.

As I worked my way through this project, I found myself realizing just how thankful I am for my life. I realized that God allowed me to fulfill a dream. I also understand what an important role the Indy 500 has played in who I am. When the book was completed, I was grateful but also a just a bit sad. You see, like A.J. Foyt, Rick Mears, the Unsers, the Andrettis, and others who have spent so many Mays at the corner of 16th Street and Georgetown Road, I don't know what life would be like without Indy in May.

If that sounds just a little too melodramatic, then I apologize. But if you've ever been to the Indianapolis 500, I think you know what I mean.

Acknowledgments

There are so many people to thank.Thank you to Tony George and the Hulman family. From day one they supported me and offered encouragement.

Most of all, I thank my dad for passing on to me his love of Indy. He passed away in 2006, and not a day goes by, that I don't miss him. Dad had six other children, but none can lay claim to the unique bond that we shared. It was ours and ours alone, and no matter the circumstances, that shared love was the way we often expressed our love for each other.

(Photo courtesy of the Indianapolis Motor Speedway.)

MY FIRST INDY

My first trip to the Indy 500 was a high school graduation present from my father, Jack.

Every year, he and his friends had made the trip from our home in Connecticut to the Indianapolis Motor Speedway. Even though my passion for Indy—which my father had instilled in me since I was five—was our connection as father and son, he never would bring me to the race. He had convinced me that children under 16 simply weren't allowed at Indy, so I listened to the race on the radio and dreamed some day of being there.

That moment came in 1969 when I was 18, and my dad finally took me. They called me Jackie back then, and I went with him and his two friends, Ray and Rich Garuti, his partners in modified stock car racing in New England.

We had silver badges that gave us access to Gasoline Alley, and I immediately recognized the sights and the sounds. There was Al Unser Sr. and Bobby Unser. Over there was J.C. Agajanian, who politely gave me an autograph.

When we walked in the gate, nothing surprised me, because for nine years, Sid Collins had described the Indianapolis 500 for me. So when I walked in, it was like, "Oh, yeah, Sid described it this way."

I soaked in every minute of this first day, snapping at least 300 pictures with my camera. Lying in bed that night at the Holiday Inn across the street, with all four men sharing a room and two to a bed, my father began to quiz me on the past history of the 500.

"Who won the 1959 race?" he asked.

"Rodger Ward!"

"Who was his sponsor?"

"Leader Card!"

"Who drove the Belond Exhaust Offy in 1957?"

"Sam Hanks and then the next year it was Jimmy Bryan!"

"Name the only three-time winners of the race?"

"Louis Meyer, Mauri Rose, Wilbur Shaw, and A.J. Foyt, but Foyt went on to win his fourth and is the all-time win leader!"

That was the kind of background that I got exposed to. I got the tribal rite handed over to me. We spent the next three days at the speedway, the nights at local tracks in the area watching smaller series compete, and I got caught up in the whole drama of what Indy means.

Finally, it was race day, and we trekked to our seats in the paddock penthouse—the same seats we still own to this day. The seats began to fill around us; the tension began to build.

At last it was time to start the race.

I remember watching the crowd fill up and my dad turning to me and saying, "There are more people here than any other sporting event in the world." All of these things I had heard about, and now I was witnessing it.

When they said, "Gentlemen, start your engines," it was a culmination of all my dreams. And when the field rumbled off, I just let all the drama of it soak in.

❖❖❖

HANDS ON THE WHEEL

The craziest guy I ever met at Indy was Jim Hurtubise, a driver with tons of promise in the late 1960s and early 1970s. I remember he showed up at a small track where my father's car raced, the Stafford Motor Speedway in Connecticut.

He had an FIA license and was looking for a ride that night in a NASCAR modified. Well, the NASCAR chief steward was a guy named Fran Grote, and when "Herk" handed him his FIA license—which allowed you to race Indy cars, NASCAR Grand Nationals, and even Formula One cars—Grote growled at him.

"Buddy, you need a NASCAR license to run here!"

Herk didn't protest. Instead he dug out $75 in cash and bought a license.

I said that Hurtubise was a talent with much promise. He'd driven a front-engined Novi to the front row of the 1963 Indy 500. But a fiery crash in June 1964 left his life in the balance.

Surgeons explained to Hurtubise that the fire had destroyed his hands to the point that all they could do was one of two things—fashion bone and skin grafts in such a manner that he could tie his shoes and accomplish other everyday tasks or mold them so he could grip a steering wheel.

With Herk there was no debate.

"Get to work, Doc, I've got races to run!"

Hurtubise didn't make the 500 field my first year there in 1969, but I made a beeline to his garage in Gasoline Alley. He was a hero to me—not because of his driving record but because of his hands.

It didn't matter who I was. When Herk saw me skulking around the front door to his garage, he crinkled up his scar-pocked face and said, "Ya wanna come in, kid?"

"Uh-huh, sure Mr. Hurtubise. Thank you, sir," I stammered.

Inside was a complete front-engined roadster. My dad said that the car was of Herk's own design and called a Mallard. It looked to me to be about twice the size of the cars that I'd gawked at moving back and forth in the rest of Gasoline Alley, but it was a page from the past that my father had so religiously brought home to me.

I ran into Jim Hurtubise on a regular basis in the years following that meeting in 1969. In fact, in 1981, I was right back in that same garage when the gun went off to end qualifications for the 500.

Jim had his trusty Mallard sitting there in the garage—just the way it looked back in 1969. But something was a little different. When the gun went off to end qualifications, I found out what the difference was: Herk opened up the engine cowling, and sitting there, instead of the Mallard engine, was an ice chest of frosty beer!

"Dig in," he smiled. "It's party time."

Hurtubise fell upon hard times in his later years, but he was always part of May at Indianapolis. He managed a car wash less than a block from the track, and you'd see him walking around the garage when he wasn't there servicing the coin machines or filling the detergent tanks at the car wash.

But as the years went by, fewer and fewer fans remembered him. I never forgot him. Jim Hurtubise was always Herk—the guy with an FIA license and a car powered by beer!

❖❖❖

REBEL WITH A CAUSE

Lee Roy Yarbrough caught my eye with his trademark Confederate flag painted on top of his helmet as the racers lined up to start the race. When the flag dropped, the cars around him spun ahead, and in a puff of smoke he was left behind as he struggled to start his car, the Jim Robbins Special. But he just sat there alone on the straightaway.

His crew ran out onto the track. They lifted the rear engine cover and swarmed around the engine, working in a frenzy to keep their car in the race. But the pace laps continued, and the mass of cars pulled farther away. Suddenly and magically, the car fired up, everyone stepped back, and Yarbrough pulled away.

With the green flag about to drop, Yarbrough sped along, frantically trying to cover lost ground.

"Man," I thought, "look at him trying catch up! This guy has a huge set of you-know-whats."

I lost sight of him after turn 2. As the cars approached turn 4, he came back into view. Yarbrough was on the outside, still behind the pack but screaming down the track. He kept trying to gain back his position, but on lap 65 his car failed him. A split header ended his day.

❖❖❖

IN THE PITS

During the race, Lloyd Ruby—one of my father's favorites—built a lead of almost an entire lap over Mario Andretti. For "Rubes," it was his race to lose.

Ruby ducked into the pits for his stop on lap 106. It was a routine pit stop for fuel and tires. All four tires had been replaced, and the crew just waited for the tank to fill.

The tank continued to fill for what seemed like an eternity. Unfortunately, Ruby wasn't about to wait—he was ready to get back out on the track. The car lurched forward with the hose still attached. We watched as methanol spilled out onto the pit stall as the fuel bladders inside the car ruptured.

The spill ended Ruby's dream of victory. The team just helped Ruby from the car, packed up their tools, and left the cursed pit.

❖❖❖

ONE AND ONLY

Mario Andretti was more than happy to claim Ruby's dream; he went on to win his first Indy.

From where we were sitting, we could see Victory Lane. It was a dream come true to see it live.

We watched the car owner, Andy Granatelli, a man of considerable bulk who was known for wearing a white suit plastered with STP logos to Indy, plant a big kiss on Mario when he got to Victory Lane. Mario looked so small—you could barely make him out from the hordes of

Car owner Andy Granatelli plants a big kiss on Mario Andretti after his win in 1969.
(Photo courtesy of the Indianapolis Motor Speedway.)

people surrounding him. But as I watched him hold the trophy and drink the milk, he became a giant in my eyes.

My father and I celebrated Mario's victory, unaware that the great driver would never win another one and we had just witnessed history. We had shared the Indianapolis 500 and Mario Andretti's victory, and when I left, I never wanted to miss another one; I wanted to watch the scene unfold before me again and again and again. That moment became a shining example of what Jim McKay meant by "the thrill of victory."

❖❖❖

FAMILY SECRET

As we left the track, my father passed on a little secret about Mario to me that I imagine he had probably heard about through his USAC connections.

Earlier in the month, Mario had wrecked his Lotus, and he was so badly burned on his face that he didn't want to be in the picture of front-row qualifiers. So he had his brother, Aldo, sit in for him.

For years, it was a secret, almost taking on the status of an urban legend—except in this case it was true.

❖❖❖

The photo of Mario that my father told me about as we left Indy.
(Photo courtesy of the Indianapolis Motor Speedway.)

THE GREAT DEBATE

Growing up, the great debate was always who is the greater driver: Mario Andretti or A.J. Foyt. If you picked Foyt, well, then you were anti-Mario. If you picked Mario, you were anti-A.J.

They polarized fans. You were for one or for the other. That was long before all of the CART rhetoric and long before the IRL existed.

But people have always figured that the common ground was Indy, and when you got into the arguments, the one thing that A.J. Foyt fans would always point to as a tie breaker was that Mario only won the race once—A.J. won it four times.

Just as quickly, the Andretti fans pointed out that A.J. Foyt never won the Formula One World Championship—something Mario did.

Mario's fans also cited his competitiveness in CART and argued he was a force much longer than Foyt was.

Racing greats A.J. Foyt (#6) and Mario Andretti battle on the track in 1969. **(Photo courtesy of the Indianapolis Motor Speedway.)**

But I was always in A.J.'s camp. I'm sure some of it stemmed from my father's admiration for Foyt. But I also thought that Foyt's record at Indy—four Indy 500 titles as a driver—put him over the top.

But when I went into TV, the debates always raged whenever the two drivers were mentioned. I once had a heated argument with Don Ohlmeyer when he was ABC's Indy 500 director back in the early 1990s.

"Foyt is so overrated," Don quipped. "He's over the hill and can't hold a candle to Mario's accomplishments."

That was Don's way. He had been around the Indy 500 since starting as a production assistant when he was just a teenager, but he never really went "inside" the sport except for a relationship that he had with Roger Penske stemming from Roger's non-racing business interests.

But because he was friends with Penske, he thought he was a racing expert.

"Uh, Don?" I sheepishly asked. "How many Indy 500s has Mario won?"

"Winning Indy shouldn't be so important!" he fired back.

"No? Why not? It's the biggest Indy car race in the world!" I snapped.

Don never liked to be challenged.

"Jack, Mario's a Formula One World Champion! Foyt never did that!"

"Foyt won the 24 hours of LeMans!" I countered.

"Andretti is more versatile!"

"Oh really? How?"

"He's won in more types of cars than A.J."

But when I pointed out to Don that he was not correct on that point, he groused, "Well, Foyt never won a Formula One race—that's real racing!"

I just shook my head and walked away.

Even after I got to know both A.J. and Mario and got to see their careers firsthand, my opinion never wavered. I have the utmost respect for Mario and his accomplishments, but Foyt is still number one in my book!

❖❖❖

A NEW GENERATION

My father stopped going to the Indy 500 a year or two after my first race, but not me. I was hooked. I have missed just three 500s since 1969.

In a way, it was almost like my father had passed the torch. I have tried for years to bring him back to the race, and sometimes he'd go to the track, but after the early 1970s, he never again went on race day.

In 1985, I was able to invite my dad to the Indy 500. He came out to the speedway and spent the week before the race with me. He got to be a fixture on pit road. By then, everyone knew me, and he would say, "That's just the cheap imitation. I'm the original!"

Almost 35 years later, I have one unfulfilled dream: My father, who is in his late 70s, always wanted to own a car in the Indy 500. Even now, I'm still looking for a way to make that happen. I feel it is something that I can and should do because, God knows, he helped me realize my dream.

❖❖❖

BREAKING FOR INDY

My Indy broadcasting break came in 1984 when my name came up at ABC as a replacement for Chris Ecknomacki. But I needed a reel, something a guy with only radio experience didn't have.

I went to the interview anyway and hit the jackpot when network officials asked me the dreaded question: "What would you do if we asked you to interview A.J. Foyt?"

At the time, everyone was terrified of A.J., but I knew him from NASCAR and thought it was a pretty stupid question.

I said, "I'd go interview him."

After covering one USAC sprint race in Rossburg, Ohio, ABC gave me my chance. They said they were going to try me out at the time trials at Indy, and if it worked out, maybe, just maybe, I could stay on for the race.

As luck would have it, A.J. Foyt was in a terrible mood when I got to Gasoline Alley. I went to his garage, and he was barking at Jack Starne, one of his mechanics. I don't remember what they were talking about, but I do remember his words were full of expletives. I thought it would be better to come back later. But as I turned to leave, he saw me. He nodded as a king would to a commoner and asked, "What are ya doing here, Jackie?"

"ABC hired me to cover the pits this weekend."

"You?"

"Yes, sir, me!" I stated proudly.

"You think you can do that?" A.J. laughed.

For a moment I was hurt. I thought maybe he thought I couldn't do the job. But then Foyt boomed, "Hey guys, lookey here, we got us a new Chris Ecknomacki!"

After the laughter died down, I told him frankly, "I really want this job."

A.J. gave me a little wink that I would come to know well. He kind of took me under his wing, and he became my ace in the hole.

❖❖❖

FOLLOWING FOYT

ABC let me stay on for the race, and I was assigned to follow A.J. Foyt on his walk to the grid. A.J. promised to be ready for the walk at exactly 10:17 a.m. But as the time neared, he was nowhere to be found.

At 10:15 a.m., producer Bob Goodrich asked, "Where's Foyt?"

Right then, A.J. came out and headed to the men's room. My whole life flashed before my eyes. I was so scared that if Foyt blew me off and failed to do the walk as scheduled, ABC would drop me from their team.

I remembered when I had interviewed for this job and ABC had asked what I would do if they asked me to interview A.J. Foyt. At the time I had boldly bragged, "I'd go interview him." Now it was "put up or shut up" time. I had to deliver Foyt, or the ABC brass would think I had been bullshitting them.

"Where's Foyt?" the crew was screaming over and over.

So I followed him to the men's room. He was already in one of the toilet stalls, which at the time didn't have doors. I didn't want A.J. to know I was following him, so once I peeked inside the john, I stepped back just behind the door to the men's room. I was out of direct view from the stalls and from A.J. I was nervously trying to figure out what I was going to do.

But then my crew started yelling at me on my headset.

"Jack," they screamed. "Where is Foyt? We need him now!"

I tried to ignore them.

"What are you doing, Jack? What's going on? Hello? Jack! Hello?" So I told them as quietly as I could, "He's in the bathroom!"

From the stall, I heard A.J.'s deep laugh. He came out and said, "Ok, Jackie, let's go."

And that was how my Indy 500 career basically started—in the men's room with A.J. Foyt.

❖❖❖

The Andretti boys—Mark, Michael, John, and Jeff—in their racing suits at Indy. **(Photo courtesy of the Indianapolis Motor Speedway.)**

FOLLOWING IN HIS FATHER'S TRACKS

Mario Andretti is quick to admit that his passion for racing probably cost him dearly when it came to his kids. He missed a lot of their childhood, leaving all of it to his wife, Dee, while he played race car driver all over the world.

"I was never home a lot. Instead it was always the next race," he told ESPN Classic when the network did a SportsCentury show on him and his career. After his retirement he was able to reflect and see the toll it had taken on his family. So his son Michael's decision to follow him into Indy cars was something that both surprised and honored Mario.

I followed Michael from his first day to his last day behind the wheel at the Indianapolis Motor Speedway. He had come up through CART's support series, and that's where I first met him and watched him race.

I remember how very young he looked when he first came to Indy. He was walking around Gasoline Alley in a firesuit, and he looked more like a kid at Halloween than a professional race car driver. He was real quiet. He had a baby face and wavy hair, and his features didn't have the character lines that millions of racing miles had placed on his dad's face. Mario's face had so many lines from so many years of racing that Jim McKay once called it "the face of a 20th-century Roman centurion."

But then Michael stepped into the car, and you had no doubt that he was going to be great. He was an Andretti through and through. He just let his driving do his talking.

His last name was Andretti, so naturally, when he expressed an interest in racing, there were immediate expectations thrust on him. When your last name is Andretti, there is an enormous amount of pressure to live up to it and do equally—or better—what Mario has done.

But Michael could hold his own. He sometimes looked like he was pushing the car beyond its normal limits. He was very aggressive, but he was always under control. He had the same kind of car control that his father showed throughout his career. Mario could see it in his son. During an interview on my ESPN radio show last year, Mario explained, "You know, all Italian fathers feel a warmth inside their chests when their children decide to follow their father's career path. But I knew how demanding racing was and how much singular focus that it took.

"When Michael was a youngster, I wasn't sure that he had that fire. But once he started racing Formula cars and the like, I knew that he was going to be good at it."

❖❖❖

INDY'S GUIDE

The nicest guy I ever met at Indy was the late Clarence Cagle, the former superintendent of the Indianapolis Motor Speedway. Clarence is credited to this day with helping Tony Hulman revitalize the Indianapolis Motor Speedway when Mr. Hulman purchased the track from Wilbur Shaw after World War II.

He was my go-to guy early in my TV career. I'd go to him with endless questions about how things worked at Indy and what he thought I could do better. He was my Yoda.

"Jack," he once told me, "Never take this place for granted. Your job here comes with a tremendous responsibility. You are responsible for insuring that the Indy 500 is always treated with the respect that is due a great race and great racing facility like Indy."

Clarence created a legacy that to this day revolves around that premise. All of the improvements in recent years—the new Gasoline Alley, the Formula One track and garages, the Bombardier Pagoda, and the fan plaza behind the pagoda—are pillars of that mantra. They are all monuments to the legacy of the Indianapolis 500. After Clarence retired from the speedway, he moved to Ormond Beach to enjoy his retirement years.

But that wasn't going to happen. When Daytona needed an efficient traffic pattern for the growing crowds that flocked there in the 1980s, Clarence signed on as an adviser and mapped out a plan that got almost 150,000 NASCAR fans in and out of the track with incredible efficiency.

When Indy needed repaving, the Hulman family called Clarence out of retirement to oversee the operation.

In the late 1990s, I asked Clarence to help my dad and my brother with their racetrack's repaving. Cagle flew to Connecticut and surveyed the 30-year-old pavement and said, "Jack, this stuff must have been pretty good ... a long time ago."

Then Cagle supervised the design and the pavement that went down at my dad's track.

It was the last pavement job that Clarence Cagle supervised. He passed away in 2003. He always refused to take a dime for his labors.

"Just remember what I told you about Indy," he would say.

I always have, and I always will, Clarence.

❖❖❖

SECRET HIDEAWAY

Michael Andretti had a terrific hideaway his rookie year. It was located on the backside of the old pit road grandstands. If you were walking out to the pits from Gasoline Alley, it was just to the left. The first area was where the pace cars were housed, and to the left of that was a helmet company's repair and prep station area.

The helmet company serviced all of the helmets worn by their drivers. The technicians made sure the hinges on the adjustable face shields were working and the interior padding was set. They were always busy.

Well, in contrast, behind the service station was a lounge with two couches and a color TV set tuned into the practice coverage on the track. Guys brought their helmets in for service and sat down to watch TV, trade stories, and discuss progress on the track. Before you knew it, the hideaway was a place for good-natured ribbing and a lot of practical jokes.

I just wandered in one day—introduced myself and sat down. There were no restrictions, just be conversational and be sure not to interrupt the drivers during their down time.

At first, I just kept my mouth shut. I was the new kid and didn't want to overstep my bounds. But after a few days, I was included in conversation, and we all began to get familiar with each other.

I remember feeling as if I had been granted admission to a baseball clubhouse. You just hung out and listened to all of the stories. I learned something every time I was there. I overheard a driver telling another driver about a conversation he had with his crew chief. Another racer talked about what was wrong with his car, and I'd take that information and keep an eye open for any progress he would make the next time out.

When that driver went out to qualify, I'd often tell some of the stories learned in that room to the TV audience—of course, without giving up any confidential information.

I mean, I wanted to be able to get back in there!

❖❖❖

ROOKIE RUN

My first rehearsal at Indy was junk. Live TV was still very new to me, and now I was no longer a spectator or fan; I was inside the Greatest Spectacle in Racing.

My biggest problem was my eyes. They kept darting all over the place. I wanted to take in everything that was

Although I am not as distracted at Indy now, there is so much going on when I am broadcasting that it brings back memories from my rookie year in pit row.
(Photo courtesy of the Indianapolis Motor Speedway.)

happening around me. The noise clamored around me. The speedway's security guards were blowing their whistles to clear paths for cars to head out into their pits. The PA was blaring, reciting for the assembling crowd the participants in the prerace parade.

Thank heaven for my cameraman, Tony Gambino. Tony was a native New Yorker and easily could have fit right in on The Sopranos.

"Yo Jack!" Tony said after I flubbed things a couple of times. "Relax. Look here into the camera and pretend that you're talkin' to da guys in ya livin' room."

Somehow I calmed down and finally got through rehearsal. That's when I said to Sam Posey, our upstairs commentator, "Sam, I finally understand why rookies crash here. It's overwhelming."

❖❖❖

SNAKE IN THE GRASS

During my first year at Indy with ABC, I was hanging out in A.J. Foyt's garage talking to one of his longtime crew members. A.J. couldn't resist the temptation to test out a practical joke. He snuck up behind me and draped a length of black hose on my shoulder. My peripheral vision caught it, and immediately my brain shouted, "Snake!" I must have jumped as high as an NBA power forward, because A.J. and his entire crew doubled over in laughter.

I, on the other hand, had to catch my breath and fight the hyperventilation that convulsed me. I am deathly afraid of snakes. From that simple rubber hose grew rubber snakes, belts, and a nonstop threat that a visit to the Foyt garage would result in some sort of reptilian encounter.

While I was interviewing Kenny Brack during one of ABC's qualifying shows, Foyt threw a rubber snake across the floor that came to rest at my feet.

"Whoa! What the ... ?" I jumped, did a midair pirouette, and kicked that rubber snake back, much to the delight of the crew, Foyt, Brack, and the millions of TV viewers who got to watch my reaction.

❖❖❖

ALL ACCESS

When I am covering a race, I feel like I have been granted access to a locker room, and the payback is that I don't purposely throw somebody under the bus.

If you see a guy who is extremely emotional and you know that he may say something that you don't think exemplifies who he is, maybe you wait 30 seconds. Or maybe you do the report with that happening over your shoulder.

But you don't ever try to put somebody in a bad light. There are an awful lot of people who say that's wrong in this industry. But I've been doing this for 20 years, and I don't regret my credo, because I think it is a privilege to have that kind of access.

I have always felt that I am the fan's witness. That's all I am. I need to bring the fan to it.

❖❖❖

PROUD PAPAS

Rick Mears is the first guy I ever interviewed in Victory Lane, and I'm not sure who was happier to be there—him or me! It was my first Indy 500 Victory Lane, and his second win, so I guess he was a bit more composed than I was.

There was a ton of activity around us. The yellow-shirted security guards were hustling around to put the Borg-Warner Trophy on the rear wing of his car, the 500 Festival Queen was smiling and posing for pictures, and the representatives of the Dairy Association were all thrusting a bottle of milk into Rick's hands.

I don't remember exactly what I said to him, but I was in complete awe of being there. I also was so relieved that it was Rick I was interviewing. He is so smooth that I knew he wouldn't let me make a fool of myself.

Near the end of the questions, I noticed his dad, Bill, was standing off to the front corner of the race car with a smile stretching ear to ear across his face. So I turned to Rick's dad and said, "What's it like to have a son who is an Indy 500 champion?"

He just teared up and started to cry. He pushed through the tears to say that he was proud of Rick—he was proud of both his sons, Rick and Roger, and was just honored to be in Indy's Victory Lane.

I hoped my dad was just as proud and smiling just as wide while watching the interview on TV. A couple of weeks later, I asked my dad if he saw the interview. He couldn't even speak. Like Rick's dad, my father teared up and just nodded his head yes.

❖❖❖

PARTY QUESTIONS

The television broadcast always ends with an interview with the winning driver. Well, when Rick Mears won his second race in 1984, he was invited up to car owner Roger Penske's suite, and it was my assignment to bird-dog Rick and to make sure he got there for the final segment.

When I finally found the three-time Indy winner, he was pretty much five sheets to the wind. He had a huge smile on his face and was sitting in Roger's trackside suite and enjoying champagne directly from the bottle. You can't blame the guy—he started celebrating at 2 p.m. when he pulled into Victory Lane, and it was now 11 p.m.

So I tried to rework the questions so that there were a lot of yes or no answers. As I escorted him to the TV position at the base of the scoring pylon, I let Jim McKay and the producers know that he was a bit inebriated.

McKay was such a professional that he too quickly made sure that he asked Rick a lot of yes and no questions. Rick performed without a hitch.

Years later in 2002, after Rick hung up his helmet, his drinking became an issue. Rick checked himself into a rehab unit for alcohol abusers. That same year Al Unser Jr. made a much more public admission. Mears went about his rehab out of the public eye and never revealed his problem until after he completed treatment.

❖❖❖

Bobby and Al Unser Sr. discuss the track.
(Photo courtesy of the Indianapolis Motor Speedway.)

THE BROTHERS UNSER

Bobby Unser always wanted to understand the race car and was always looking for a competitive edge. I remember one day when he was huddled over with the folks at Goodyear trying to come up with the idea of putting crushed walnuts in race tires for his tires to run at Pikes Peak. Bobby was always the guy who attacked the mechanical side of things.

Bobby was a race car driver who always fancied himself better than most engineers. He was always coming up with ideas about the car. One time he suggested that his crew send the car out without its rear wing.

"I'm telling you," he whined. "That big ol' wing is slowing me down. Just take it off and the car will go faster."

"But Bobby," the engineer protested, "without the rear wing, the car will lose rear downforce, and when you hit the turn, you will either crash or have to hit the brakes."

"Why don't you let me worry about that, cousin? I'm the driver," Bobby said.

They never did remove the rear wing. But I'm sure sometime during a private test session, Bobby got his way and got out on a racetrack without a rear wing.

Al Unser Sr. was a guy who wanted to sit in, strap in, shut up, and go racing. He certainly had a lot less to say about things than Bobby did, but the competitive fire was exactly the same. He didn't want to be part of a developmental program. He was a race car driver; he wasn't an engineer. But as Indy cars progressed, so did their sophistication.

Al once told me that a lot of the telemetry and gauges that found their way into the cockpit often distracted him from his driving.

"One time," he told me, "one of my mechanics decided that he needed a running commentary during the race."

This was before the cars carried the computer systems that automatically transmit data back to the pits. So the only way to get the information was over the radio.

Al said the mechanic peppered him with questions while he was out there running.

"What's the oil pressure? How 'bout the fuel pressure?"

Al kept trying to ignore all of the radio inquiries, but eventually he had to answer.

"Al, what's the water temperature?"

"I looked down while tooling down the backstretch to try and read the water temperature gauge, and bam, I'm in the wall!" he recalled.

Al never has liked distractions.

❖❖❖

RACING PENSKE-STYLE

Known as a smooth operator and the master of the Brickyard, Rick Mears was really a blue-collar boy from Bakersfield, California, in whom Roger Penske found a steely confidence. Roger didn't force Rick to become slick. My father always said, "If you walk with a guy who limps long enough, you begin to limp yourself." So by Roger's example, Rick saw what the advantage was.

The one thing Roger saw in Rick—Roger used to call it his desert eyes. Rick had an incredible ability of being just as proactive as Roger was from his years of running the Baja and running off-road racing, where he could see down the road only just so far. He was always able to see way down the track. He could anticipate things. When he raced to his first Indy 500 win in 1979, it was his desert eyes that allowed him to see potential race traffic well before it became a factor. It allowed him to choose the right moment to pass the lapped car and still keep the lead.

Rick was smooth. And Roger liked that. Together, they used the 1980s to climb to the top, and they've never gotten off of it.

Danny Sullivan was brought aboard to even out the balance on a team that already boasted the calm and cool Rick Mears. He was a slick, young *Miami Vice*-type driver whom Roger brought on board to complement or to be the yin to the yang of Mears.

Rick was slick in the cockpit but quiet and reserved out of a race car. And when it came to schmoozing VIPs, Rick was always courteous, but he was far more at home talking race cars with mechanics and engineers than with the heads of Fortune 500 companies. Penske needed someone to parade in front of his VIPs and financial supporters who was adept at cocktail talk. That was Danny Sullivan.

Danny was all about image. He had a full-time Hollywood press agent named Alan Nierab who made sure that Danny was in *People* magazine and invited to all of Hollywood's "A-List" parties ... before he ever won the Indy 500!

But Sullivan was scared to death on race day, something I could sense on pit road just by looking at him. I remember standing with Danny the day of the race, just before we went on the air. He looked very nice in his flame-red Miller High Life driving suit. But when you looked Danny Sullivan in the eye, you knew that he knew he was in over his head. What he said to me, though, was that he knew that he was driving for the best team in the world and he said he knew "That will get me through the rough spots." And that's the one thing that drivers who raced for Roger Penske have always known.

❖❖❖

SAFETY FIRE

Few people have done as much for racing as Bill Simpson. He started out as a driver but quickly decided that his second career should be devoted to increasing the safety of drivers.

He founded a safety equipment manufacturing company that to this day bears his name even though he has sold it and formed a new one called Impact Racing.

Simpson is a man who speaks his mind. He's a no-bull kind of guy who demands results. Back in the early 1980s, there were a lot of drivers who paid little attention to what safety equipment they were using, especially driving suits. So Simpson hooked up with some engineers at Dupont and started manufacturing suits made out of Nomex fibers that were far superior to some of the other products on the market. He wanted a way to get attention for his suits, so he sent out a letter delivered to every garage in Gasoline Alley and to the press room at the Indianapolis Motor Speedway.

He challenged other suit manufacturers to a side-by-side comparison later that day. He said he would don his Simpson gear and set himself on fire. He wanted the other company reps to do the same.

"We will see who's ready to put their ass on the line," he growled.

No one took Bill up on his offer, but that didn't stop Simpson.

Surrounded by a small crowd of media, drivers, and crew members, Simpson strutted out in a triple-layer Simpson firesuit complete with underwear. He had on one of his Indy helmets with a hood that closed the area between his helmet and the firesuit. An assistant doused him in fuel and lit him up.

Whoosh! Bill Simpson was on fire! Just like in the movies.

But this was no stunt man. This was a guy who was trying to prove that his stuff worked!

It did, and by the next morning, the performance was the talk of Gasoline Alley.

❖❖❖

SCHOOLS OF RACING

A lot of people want to make Michael Andretti a clone of Mario, but he's not. They really come from two different modes of racing. Mario was old school. Mario didn't become close to other drivers. He knew that racing was dangerous, and any given week they could die on the track. He also knew that in the race cars, everyone else was out to beat him to the finish line and take potential winnings that he needed to feed his family.

Michael came from the next class, where racing had become more of a fraternity of brothers who share and interact with one another. He was more aware of his surroundings. With his helmet off, he loved to chat and joke with other drivers.

I knew right away that Michael was different when I saw the friendship that he had with Kevin Cogan when they both first arrived at the speedway driving Maurice Kraines's KRACO cars. They hung out together and talked about more than just how the car was running: girls, good places to eat, and events in the sports world outside of Indy.

Michael was and still is a big hockey fan. Ask him what he thinks of Eric Lindros or one of the Philadelphia Flyers' lines, and he will jump right in with an opinion. It's always part athlete talk and part fan talk.

As serious as he was about racing, Michael also had a terrific sense of humor.

Drivers' meetings can be a little tense. Especially the private one held prior to the Indy 500. Indy holds a public drivers' meeting, where all 33 guys sit on a stand and fans snap photos, sponsors get recognized, and special awards are distributed, but it's the private meeting where the nitty-gritty gets sorted out.

Well, one time, Indy's chief steward was going over some rules for the race, and the conversation was getting heated. Drivers were debating USAC's decision to penalize those who put their wheels below the white line that had been painted on the interior side of the track about a foot up from the edge of the pavement.

Michael was right in the middle of the debate because he had perfected—as had his father, Mario—the ability to run very quick below that white line. Now, just before the race, USAC was going to take that line away.

"Why have the blacktop there if you can't race on it?" asked Michael.

"It's a safety issue," was the reply.

That's when Michael quipped, "Heck, it's a lot safer down there than up where everybody else is racing!"

It cracked up everyone in the room.

But don't think for a minute that Michael was any less intense than Mario on the track. When Michael would put the helmet on, he would wreck you as soon as race you. There were no holds barred, just like dad.

❖❖❖

SUITED FOR THE ROAD

When the powers that be required firesuits on pit road back in the late 1980s, Roger Penske showed his commanding sense of style.

Penske thought it was unbecoming for someone of his stature to be dressed in a flame-resistant Nomex jumpsuit. So he went out and had a Nomex shirt and Nomex black slacks made so that they looked like a normal, Penske-esque pit crew uniform. It wasn't flamboyancy. He

Roger Penske (left) always dressed the part.
(Photo courtesy of the Indianapolis Motor Speedway.)

did it so that until you got up close, you thought it was a normal pair of slacks and a normal shirt. Then you realized it was Nomex. He just oozes with class like that.

❖❖❖

AT A LOSS

When A.J. Foyt lost his father, Tony, it had such a negative impact on A.J. that it took him a number of years to recover sufficiently to get back to the business of racing.

Tony kept A.J. grounded when he raced. He was the only guy who was fearless in A.J.'s presence, and I think that it was Tony Foyt's input that provided the balance and focus he needed to win four Indy 500s.

Like a lot of Texans I know, A.J. is impetuous, prone to outbursts of anger, and sometimes can let his emotions override his common sense. Well, when his daddy was alive and that happened, his father would literally take him outside the garage and slap sense into him.

Immediately following his father's death in 1983, when A.J. was at Indy, you could see that he felt the absence of his father. There were times when he'd break out of his funk and display flashes of his old, volatile personality, but there was sadness to him. I knew the relationship he and his father had and wished Tony was there to remind A.J. that he was tougher than that.

Shortly after Tony's death, A.J. invited me to his ranch, and we spent hours reminiscing about his 30 years at Indy. Seldom was there a sentence he uttered that did not include a reference to Tony. A.J. took me to his father's house and over to a little garage that Tony always used for

A.J. Foyt is a true Texan. Notice the boots.
(Photo courtesy of the Indianapolis Motor Speedway.)

projects that he wanted to tinker on away from others. In the garage was a toolbox that his father had used from when he started as a midget owner back in the 1940s until his death. A.J. spent an hour just showing me each and every tool in that box. Each was hand-engraved with Tony Foyt's name or initials, and each had a story attached.

"This is the wrench my daddy used in 1953, and this is the wrench my daddy used when we first won Indy."

It was an eerie show and tell. This was A.J. Foyt caressing the last thing that connected him to his dad. I felt uncomfortable standing there, but as A.J. went through the toolbox, you saw both the sadness and the pride that A.J. possessed. That was when I first realized that until A.J. could have closure with the loss of his father, he would no longer be the great A.J. Foyt I knew.

TIME TRIALS

When you work in television, you lose all track of time. Everything that happens in the month of May is based upon track activity.

What time is your production meeting? Five minutes after Happy Hour ends. What time are you going to meet with the officials? Just before practice starts. What time are you going to do the interview? At the lunch break.

It's not a number on the clock; it's all based on track activity, and you get caught up in that. Some drivers leave to refocus. TV guys don't get a chance to do that. So by the time you get to race day, you are both exhausted and all hyped up.

When I first went there, I remember trying to figure out why rookies made mistakes. You look up, you see the crowd, you know that you are about ready to go on the air, and you get caught up in everything in and around the event. No matter how calm you try to be, it does affect your personality and the way you are going about your job. The rush is there. It's only after a few years of covering Indy that you discover the ways to keep it in check. When you are in the midst of it, you are a part of it in a small way.

I've always believed that part of the mad rush is because up until that day there wasn't the crowd. Every day there wasn't the electricity, there wasn't the Purdue Marching Band, and there wasn't the parade of stars.

When you throw all of that into the mix, all of a sudden it starts an adrenaline rush, and you've got to keep control of that.

❖❖❖

PERFECTLY CURSED

Mario Andretti was driving a perfect, masterful race at Indy in 1985. He was on his way to his second Indy 500 win, and that's when fate or the curse—or whatever you want to call it—stepped in.

Mario and Danny Sullivan were trading the lead back and forth. Mario was hounding "The Hollywood Kid" just in time to see Danny spin. I mean, he did a complete 360!

Anyone else would have hit the wall in that short chute between turns 1 and 2 and ended his day. But not Danny. He didn't hit a thing, ducked onto pit road, and eventually beat Mario to the checkered flag. He won his first and only Indianapolis 500.

I don't know what more Mario could have done. He drove the perfect race. He had led 107 laps of the race, ran well in the pits, and was never out of the front pack of cars. Still, he finished second.

We in the ABC crew had been sure that Mario was going to win. I remember during commercial breaks, most of the talk between the announcers was marveling at how well Mario was doing. When he won his first race in 1969, everyone thought he was starting out on a career that would include at least a couple of Indy 500 wins.

I walked up to Mario after the race. He was spent. He'd just driven 500 miles, and his red driving suit was soaked through with perspiration.

"What do you think you have to do to win this race again?" I asked.

I could tell immediately how frustrated he was, because he barked at me and denied that fate had anything to do with Danny Sullivan beating him. It was one of the more telling moments for me in my career on pit road, because Mario Andretti wasn't trying to convince me—he was trying to convince himself.

That was when I began believing in the Andretti Curse.

I walked away thinking: "I hope Mario can get that monkey off his back and win soon, or he will be a basket case."

❖❖❖

TWO-FACED

Rick Mears once told me that one of the things that auto racing afforded him was the ability to be somebody he's not.

What he meant was that he could talk, do interviews, be around folks, have a wonderful time, and have a nice calm demeanor outside a race car. When they would go testing, he could be analytical: He could tell you what the car was doing, and he could get all into the nuances and the art of driving a race car. But when it was race day and it was time, he would put that helmet on and it was like Dr. Jekyll became Mr. Hyde.

He could become a stark, raving lunatic in the race car. Rick said that when he put the helmet on, he became someone who he himself would not want to hang out with. But when the helmet came off, that personality went with the helmet back into the bag. He's one of the few guys who I have seen that was able to successfully do that through an entire career.

❖❖❖

A SPECIAL DAY

Every year at Indianapolis, Mari Hulman George holds a special picnic for mentally and physically challenged kids. It happens on the first weekend of practice and always reminds me of what really matters. It's called Save Arnold Day and was given that name many years ago.

Kids from all walks of life and situations flock to the backside of the Pagoda and get a chance to play baseball and other games with all of the Indy car drivers. These kids are just thrilled to be there, and you watch as they get autographs and talk to the drivers. You realize that to them each and every one of these drivers is a hero. It doesn't matter to these kids whether the driver they are talking to has ever won the Indy 500. They don't care if the driver is a rookie or if he even has a chance at making the 500. They are just excited that these stars have taken time to visit with them.

The impact is far greater upon the drivers and crew members who participate in this annual affair. The late Tim Richmond once told me that Save Arnold Day made everything that happens in the month at Indy palatable.

"Jackie," he said, "these kids will make you forget the worst of times here."

Rick Mears in action on the track.
(Photo courtesy of the Indianapolis Motor Speedway.)

It's always fun to watch these athletes swing a softball bat or compete in a relay race. They may be athletes in their own sport, but some of them really make you laugh when they step out and try to hit a ball or sink a basket. I think that delivers a positive message to the kids. Although these guys are outstanding race drivers, they are no different from most guys on the street.

They are better than most, though, because they take the time to be part of Save Arnold Day.

❖❖❖

MAY MADNESS

There was a time when the entire month of May was Indy for drivers. Every day was devoted to practice, and it was easy to lose track of what was happening outside of 16th Street and Georgetown Road as they tried to crack Indy's code.

After qualifying for the race, Bobby Rahal told me, "I have to get out of here." And he got his stuff and left. He was gone for a couple of days because he needed to get his head screwed back on straight. He didn't even go very far—just back home to Ohio. Once away from Indy, he hit the golf course and played a few rounds. He relaxed at home, played some more golf, and relaxed a little more. He did anything and everything so that he wouldn't think about Indy, what was going on at the racetrack, and what he had to do when he got back.

You see, Indy plays tricks on the mind. And if you're a driver, you've got to have a clear focus and a clear mind. Rahal knew that, and when he came back, it paid off.

❖❖❖

STUDYING FOR SUCCESS

Rick Mears always studied the styles of the drivers he ran against. When he would come up on a driver, he used that knowledge to set up and pass the guy with a move that just looked effortless. He was so good at qualifying. In 1986, he was practicing for qualifications. Whenever you went by Rick's stall in Roger Penske's garage, Rick would be huddled with his mechanics and engineers. There was no look of concern. Instead it was like a group of guys trying to solve an arithmetic equation.

"I think if we change the front wing angle, this will happen."

"We could narrow the track of the car, and this will happen."

They tinkered and adjusted. By the time Rick rolled out onto the track, all of the changes were made, and voila! Rick Mears earned the pole for the 1986 Indy 500!

❖❖❖

CAUTION TALK

The rush from the pressure of a television broadcast is indescribable. If you are sitting at home watching and wondering, "Why did they say that?" Well, we're not doing it because we're dumb. We're doing it because we're trying to bring people where they can't go.

In 1986, everyone decided they were going to try to talk to drivers under caution. During the race there was a late caution, and Kevin Cogan was leading the race. Sam Posey decided maybe this would be a good time to talk to Kevin. Well, by the time they finally got around to doing it, there was one lap to go before green. So Sam got on the radio and said, "Kevin, this is Sam up in the ABC booth. You are having a wonderful drive today."

And the classic line from Kevin was, "I can't talk to you right now, Sam. I am about to get very busy."

The race went green, and as the cars screamed past the start/finish line, Bobby Rahal dove to the inside and edged Cogan from the front spot by the time the cars hit turn 1. Rahal took that lead all of the way to the checkered flag and his first Indy victory.

After he race, Kevin was completely worn out and obviously disappointed. But he never blamed ABC. Some Kevin Cogan fans felt that ABC cost him the race. Some sportswriters even agreed with the sentiment. It really gained momentum when some TV critics recounted the story and tried to bash the network for interrupting Cogan at a very bad moment.

Kevin will still tell you that it is not true.

Even years later, when a reporter asks him to relive the situation, he still doesn't complain.

❖❖❖

REASON TO WIN

In 1986, Bobby Rahal and Jim Trueman won their only 500. Rahal took advantage of a late race restart to pass race leader Kevin Cogan and drive his Truesports car into Indy history. When he got to Victory Lane, the car owner, Jim Trueman, was there—but just barely.

Trueman was a strapping guy who had been battling cancer. There he stood, just a shadow of himself. He was maybe 100 pounds. Everybody knew he was dying; in fact, many of us wondered if Jim would still be alive for the race after officials had announced that for the first time in track

Bobby Rahal celebrates victory in 1986. **(AP/WWP)**

history the race would be postponed to the following weekend because of rain.

But there he was with a frail smile on his face. We did the interview with Rahal and Trueman. Trueman had his left hand on my neck to steady himself on his feet.

Right in the middle of the interview, Bobby leaned over to Trueman and said, "This is for you, Jim."

When the on-camera interview ended, Trueman told me, "I can go now." Twelve days later he was dead.

People can survive terminal illnesses until they reach their point of closure. Jim's moment was winning Indy. He willed himself to stay alive until he got what he wanted—an Indy 500 win.

As much as Rahal won the Indy 500, it was Jim Trueman's win. No one will ever convince me that fame, fortune, God, or whatever didn't look down and say this win was for Jim Trueman.

❖❖❖

TALKIN' SHOP

One day at the speedway I was stopped by a fan, and he just wanted to talk racing.

"I was at the Speedrome last night, Jack, and saw you race your Legend car."

The Speedrome is a tiny track on the outskirts of Indy that hosts Figure 8, Midget, and Legends car races. I had towed my Legend car to Indy for some rest and relaxation.

"You did? Gee, that's great," I replied.

"Yes, it was, Jack," said the fan. "That's why I wanted to look you up today."

I figured this guy was going to compliment me on my driving prowess. After all, I finished fifth in the race at the Speedrome.

But I was so wrong!

"Do us all a favor, will ya, Jack?" he snorted. "Don't quit your day job!"

❖❖❖

CREATURE OF COMFORT

Bobby Unser literally got out of his race car and into an announcer's booth.

At Indy, Bobby covered turn 2. That's a remote location, and when the booth is being put together, very little thought is put into creature comforts.

But not for Bobby.

On race day, Bobby had a recliner, refreshments, five hangers-on, shade and cover from the wind and the rain, and all of the monitors that he needed. He was over there in turn 2 calling the race like he was in his living room!

The rest of us always tried to kid him about his setup, and as many times as we brought it up, he never let it bother him. Instead, he grinned like a Cheshire cat and said, "Son, you gotta be comfortable to do your best work!"

What was amazing was how he got his elaborate setups. Bobby had so many friends working at the speedway that he could procure almost anything. He just had so many contacts that all he did was dial someone up and start a conversation and the next thing you know he said, "Son, I realllly neeeeed a favor..."

By that time everyone was ready to do anything for Bobby. After all, we are talking about an Indy 500 winner! People wanted to tell their friends that they were friendly with Bobby and that they helped him!

❖❖❖

THE SIRENS OF INDY

If you ever get the chance, and I've done it, visit the track when it's empty. The best time is at sunset, but almost any time will do.

Then take a stroll on the track down the front straightaway. If you listen carefully, you'll hear Sid Collins welcoming you to the Greatest Spectacle in Racing. You'll hear in the silent echoes Freddie Agabashian talking about his work with the Champion Spark Plug Company.

When you cross the finish line, you will notice that the yard of bricks is indeed a three-foot section of bricks! You'll also hear Eddie Cheever Jr. telling you what his father told him early in his career, "If you only win one race, son, make sure that it's the Indianapolis 500." You will be able to hear Tom Carnegie booming out his trademark, "And heeee's on it!"

Drivers also are not immune to Indy's siren songs. Tom Sneva told me a story about his first Indy 500. He

The first Indianapolis 500, held in 1911, gets off to a smoky start. (PHOTO COURTESY OF THE INDIANAPOLIS MOTOR SPEEDWAY.)

Tony Hulman (left) and Indiana Governor Matthew Welsh place a golden brick in Indy's famous yard of bricks in honor of the race's 50th anniversary.
(PHOTO COURTESY OF THE INDIANAPOLIS MOTOR SPEEDWAY.)

said he put in miles of practice and qualified for the race when the track was empty of fans.

But he was totally unprepared for what awaited him on race day.

All of a sudden the empty grandstands were full. Smoke was wafting up from infield fans tailgating. And instead of the silence and the whispers of history, the noise permeated everything. He said, "It's a totally different world than the one you have lived in for the month, and it takes a bit of adjustment."

❖❖❖

LADIES FIRST

There have been three women who have driven at Indy during my time there: Two I covered for ABC and the other—the first female Indy driver—I watched race and met several years after she stepped out of the cockpit.

Janet Guthrie raced there in 1977, 1978, and 1979. She first arrived at Indy in 1976 but failed to make the show that year, driving for Rolla Vollstedt's Bryant Heating & Cooling car.

In 1976, Guthrie arrived at Indy and was subjected to the most scrutiny that any rookie ever has received there. She answered all of the questions and performed better than many rookies before her. By the final day of qualifying, she and her car were just not fast enough to make the field that year.

But her efforts won the heart of the speedway's Top Dog—and arguably its biggest chauvinist—A.J. Foyt.

Just minutes before the end of final qualifications, Foyt ordered his crew to roll out his backup Coyote car and told Guthrie to get in and shake it down.

After a couple of quick laps, Janet was already turning laps around 180 mph. The Guthrie/Foyt combo was quick enough to make the show but did not attempt qualification.

"I just wanted to show everyone that she could drive," Foyt told me years later.

In 1977, Janet did make the show, qualifying 26th at 188.400 mph, and became the first female driver to start the Indy 500. Her presence forever changed the command from "Gentlemen, start your engines!" to "Ladies and Gentlemen, start your engines!" Well, not quite. At least not that first year.

Janet Guthrie climbs into her race car in 1978. She was the first woman to run in the Indy 500 in 1977. (AP/WWP)

At that race, the debate swirled around how Tony Hulman would change his traditional start command to accommodate Guthrie's inclusion. In the end, he simply said, "In company with the first lady to qualify for the Indianapolis 500, gentlemen, start your engines!"

Guthrie made three consecutive Indy 500 appearances, finishing 29th in 1977, ninth in 1978, and 34th in 1979—the year that the field was expanded to 35 starters.

Vollstedt, more than Guthrie, should get the credit for breaking down Indy's gender barrier. In 1987, I went out to Colorado and met with Guthrie on the occasion of the 10th anniversary of her becoming the first woman to race in the 500.

She was living a quiet life in Aspen, and I told her straight up that I thought her accomplishments at Indy were helped along by the wave of feminism that swept the country back in 1976.

"If there wasn't bra burning and the call for female equality, Janet, you would have never made it to Indy," I boldly told her.

"You are absolutely correct," she calmly responded.

"But you need to remember that I was like every other aspiring racer. I wanted to drive at Indy, and Rolla called me. I didn't call him."

Janet had solid road racing credentials. She was a veteran road racer and knew when she got the call from Rolla that it was an opportunity she couldn't refuse.

"It wasn't about whether I was ready for Indy or whether I was a man or a woman as far as I was concerned," she said.

"Jack, if you were a young racer and someone called you and said, 'I've got a ride for you for the Indy 500, do you want it?' what would you say?"

Advantage Guthrie! She had me. And with that I became both a fan of her career and a follower.

❖❖❖

A HELPING HAND

In 1987, Al Unser Sr., an aging veteran, had three titles and no ride.

Roger Penske had a full stable, and he couldn't stand to see the driving great sidelined, so he promised Al that if he got his cars into the race in the first week of qualifying, which he did, he would give Al a car to try to get into the field.

The race car Penske gave Unser was a show car. On May 1 it was sitting in the lobby of a Sheraton Hotel in Scranton, Pennsylvania. Penske sent his crew with a rental truck, loaded the car in, hustled it to Indianapolis, and put the setups on it.

Al went on to qualify and win an unprecedented fourth Indy 500.

He had just plodded along that day; most of the race belonged to Mario Andretti. Mario totally dominated that day, leading 170 of the 200 laps only to have the infamous Andretti curse strike again when his ignition failed with just 20 laps to go in the race.

Al took over the lead and led the last 18 laps.

Al Unser Sr. races the Penske show car to victory.
(PHOTO COURTESY OF THE INDIANAPOLIS MOTOR SPEEDWAY.)

Penske had put Al in the field, and the good deed paid him back. For Al, I think that was a special victory because his son was in the race.

After the race, as Al Sr. stood there for all of the photos with the plethora of manufacturers' hats propped on his head, Little Al ducked his head into Victory Lane and sheepishly caught his dad's attention.

There was a quick embrace and a couple of whispered moments, and Al Jr. was gone. The two celebrated together in Victory Lane long after the TV cameras were gone.

❖❖❖

FIRE ALARM?

The most dangerous part of pit road is the threat of fire. The cars run on methanol, which burns a clear flame. It is very difficult to see. Instead, you feel it. The heat is incredible, and it consumes oxygen so quickly that when you get close to a methanol fire or see methanol burning, the air gets thinner and it gets increasingly difficult to breathe.

I was in A.J. Foyt's pit when the team was venting his race car. Somehow the methanol fumes ignited. The small fire spread throughout the pit. Some of the methanol from the car got on my firesuit, and it started to burn. Everything happened so fast that the only thing I could think to do was yell.

"I'm on fire! I'm on fire!" I shouted to ABC producer Mike Pearl.

Unfortunately his concern did not match mine.

"I'll be with you in a minute," he replied.

I was dumbfounded by his indifference to my situation, but in the heat of the moment—no pun intended—I just kept on reporting and seeking out the story. But the fire had to be put out, so water was poured all over me. Water is the quickest and most effective method of extinguishing methanol fires.

Luckily, I wasn't hurt and was able to finish the day. But the legs on my firesuit were singed.

❖❖❖

LIKE FATHER, LIKE SON

It was obvious from the beginning that Michael was going to be as good as, if not better than, his father.

What snuck up on him was the Andretti luck. Like his father, Michael suffered the weirdest things that would keep him out of Victory Lane.

One Indy, Michael had driven a terrific race all day and was leading with just a few laps to go. ABC had an onboard camera on his car. Sam Posey was explaining how Michael was doing a terrific job when right at that moment the side pod flew off his car!

And then in 1995, Michael qualified fourth, right behind Scott Brayton, who had won the pole. Stan Fox violently crashed on the first lap, and when the race went back to green, it was Michael out front.

He was flogging his Ford Cosworth, and it was responding with lightning quick times. He led 45 laps, but by lap 77, Michael found himself in the pits with suspension failure and out of the 500.

The closest taste of victory Michael got came in 1992. Michael and his Newman-Haas team were clearly one of the top contenders for the 500 that year. He'd qualified on the outside of the second row. He led all but 29 of the 189 laps he ran that year.

But as had become the norm at Indy, with 11 laps to go, you heard Tom Carnegie say, "Andretti is slowing! Andretti is slowing!"

His Ford Cosworth engine mysteriously lost all fuel pressure, and Michael coasted to a stop on lap 189. He had to watch as Al Unser Jr. and Scott Goodyear raced to the closest finish in Indy 500 history.

The curse followed the Andrettis to the next generation: Until Michael retired, he was the active driver who had led more races—seven—and more laps—426—than any driver not to win the Indianapolis 500.

❖❖❖

NECK AND NECK

There have been some close finishes at Indy. In 1989 Al Unser Jr. and Emerson Fittipaldi were slicing and dicing for the lead with just a handful of laps left.

Because of the way Indy is built, even the crowd can't see the entire track during the race. From my station on pit road, I see about five seconds as the cars speed past. But I could tell something was happening. As the two racers continued to stalk each other and the lap count got higher, the buzz from the crowd became louder and louder, and the voices of announcers Paul Page and Bobby Unser got higher and higher.

Little Al hits the wall as Emerson Fittipaldi speeds by to win.
(AP/WWP)

It's odd that when you are deprived of your visual senses, your other senses compensate. I just knew that something was going to happen. It was like a small charge of electricity—the kind you feel when you accidentally touch both ends of a nine-volt battery.

As loud as the cars and the crowd were, there was a stillness that I felt. And there was nothing to do on pit road except wait.

I moved to Victory Lane to prepare for the postrace broadcast.

On the track, Emo and Little Al headed into turn 3, and Al Jr. told me later, "I knew two of us were going in, one of us was coming out and the other guy might be dead. It is the Indy 500, so I'm going for it."

I heard the crowd gasp in surprise, and Paul Page said, "They touch." The noise from the crowd continued and drowned out the announcers' voices. I didn't know what had happened, who had crashed, if anyone was hurt, or who had won. Then I saw the Marlboro red race car as it scrubbed off speed, and I knew Emo had won the race. Little Al wrecked into the wall.

To many people's surprise, Little Al saluted his victorious adversary. A lot of people thought he was going to take a shot at Emerson, but when the TV showed him, he was clapping and applauding. Al's actions didn't surprise me. I knew that he appreciated that kind of flat-out racing from his sprint car days.

❖❖❖

THE PEOPLE'S CHAMP

I call Rick Mears the "People's Champion" because he never accepted his success at Indianapolis with any sort of cavalier attitude. He always loved every minute of it.

He was a go-to-work guy who would go have a Budweiser with the fans after the race and sit there, talk, and enjoy himself.

He also enjoyed himself in Victory Lane. Every time fans would start to cheer, he would look out across the expanse and just take in the cheering—his smile broadening, and then he'd turn to his crew and laugh and smile more.

Indy's Victory Lane requires a lengthy hat dance, a practice where the winning driver poses with an assortment of contingency award sponsors' hats for a commemorative photo that these sponsors can use in their advertising or just frame and put in some bigwig's office.

A lot of drivers get tired of the hat dance and just want it all to be over so that they can finish up with the photos, do the mandatory press conference, get back to the garage, and change into their civilian clothes.

Not Rick. He posed for every sponsor, smiled sincerely and obligingly, and kept smiling for more than the 45 minutes that it took to get all of the hats and photos done.

Rick loved his fans. Whenever he'd walk out to pit road, he always took time to sign autographs and recognize the fans who were wishing him well.

And he was always very, very tight with his crew. A lot of drivers would converse with their crew and then take off for a motor home or a place of refuge away from the garage. Not Rick. You could always either find him in Roger Penske's garage or back in his room at the Speedway Motel located just outside turn 2 of the track.

When daily practice ended, most drivers would hightail it out of Gasoline Alley for a personal appearance or a quiet dinner. I can't tell you the number of times I'd swing by his garage a couple of hours after practice ended only to find Rick still in the garage watching TV and sipping on a beer.

❖❖❖

THE BETTING ALTAR

Like me, Al Unser Sr. is a veteran of many marriages. We were both single at the same time and decided we just weren't cut out for marriage. So we made a bet: Whoever got married first would have to pay the other $1,000. It was supposed to prevent us from making the ongoing mistake of marrying the wrong people.

Well, I lost.

I sent out the announcements, and the first phone call I got was from Al. He reminded me that I owed him $1,000, but instead of a lump sum, he wanted me to pay him $20 every time I saw him. Sure enough, every weekend Al found me at the racetrack and asked for his $20. It was a well-kept secret that very few people knew about. Until I was on the air interviewing Al Sr. during a pregame show, and he said, "Where's my $20?" He wanted to be paid on the air! So I handed him $20 right then and there.

On my earpiece, the producer started screaming for an explanation. The viewers didn't get it, and he wanted me to explain it to my audience.

So much for keeping it a secret. I faced the music and explained the $20. Now everyone knows about the bet.

❖❖❖

TURNING IN

In 1991 I was down in turn 1 at Indy when Rick Mears had his big accident and finally decided maybe it was time to hang his helmet up.

I was watching the cars practicing, when all of a sudden I heard a "Bam!" like a loud explosion. I turned around to face the front stretch, and Rick Mears's car was sliding and bumping against the outside concrete wall.

The car was in pretty rough shape. The right side pod was torn off, and the suspension was hanging out, bent, and without the wheel attached. It was bad, but not terrifyingly bad. I had seen cars wreck a lot worse after hitting Indy's concrete wall.

Rick got out of the car on his own. I was too far away to really see the look on his face, but he did walk around surveying the damage to the car before he got in the ambulance for the mandatory trip to the infield med center.

As Rick was being examined in the infield care center, the Penske guys were trying to figure out what had happened. They called over to the ABC-TV compound looking for video of the wreck, only to learn the few shots that were available were inconclusive.

No one ever figured out what caused Rick to wreck, and I think that was because he wasn't able to go back and reconstruct it.

It scared him.

Now Rick was once in a terrible accident in Montreal when his feet and legs were destroyed under a piece of guardrail. But he battled back and got back in a car even though his feet to this day are so tender that he can't wear real shoes. He has to wear a tennis shoe with a soft insole to protect his feet.

After Montreal, he didn't think twice about getting back in the cockpit.

After the crash at Indy, Rick got in his backup car and raced just as he always did. He won pole and went on to nab his fourth Indy victory.

But I believe that when he crashed in the corner at Indianapolis and couldn't square with himself what happened, he began to wonder if it was him or if it was the race car.

❖❖❖

CAUSE CHAMPION

By the time Lyn St. James arrived on the Indy scene in 1992, the situation surrounding women behind the wheel was vastly different from what it had been when Janet Guthrie raced in the 1970s.

I first met Lyn St. James when she was driving prototype sports cars in the IMSA series. A veteran of Daytona's 24-hour race, St. James picked up IMSA's most improved driver award in the 1980s and in her acceptance speech gushed the same way Sally Field did when she won her Oscar.

"This means you guys like me! You really like me!" she said.

Our paths crossed often after that, and we always shared a dream about Indy. She wanted to get there in the worst way. Lyn peppered me with questions:

Did I know any owners who might give her a chance? What did she have to do to get a test ride in an Indy car?

Unlike the first female driver at Indy, Janet Guthrie, Lyn fully intended to use her gender as her Indy launching pad. All of her efforts were directed that way, and there wasn't an Indy appearance that she'd turn down, regardless of the appearance fee. The idea was to network and hope it led to a ride.

It did in 1992 when she came to Indy driving the JC Penney Spirit of the American Woman Lola. She quickly passed her rookie test and made the field in the 27th spot. She went on to an 11th-place finish and earned the race's Rookie of the Year honors, becoming the oldest driver to do so at the age of 45.

She made seven starts through 2000 and never finished better than that first year.

But St. James was never really accepted at Indy. It wasn't that she didn't have the credentials; she had decent equipment and could drive. There always was a prevailing feeling that Lyn played the gender card too often.

She was a former president of the Women's Sports Foundation and championed women's causes. Lyn St. James reversed Guthrie's priority order. Hers was gender first and driving second.

During an interview, you could always count on St. James to bring up the topic of gender.

"Lyn, you turned 225.340 in qualifying. Congratulations, that should be good enough to make the race," I said to her in 1995.

"I hope so," she said smiling, pausing just a second before launching into her campaign speech. "That makes me the fastest woman on wheels, and I'm very proud of that, Jack. You know, I have started a foundation to develop female drivers and open this sport up to young women."

Blah, blah, blah. It just went on and on. In Lyn's defense, I should say we at ABC played the female angle to the hilt.

She was always featured on our telecasts, not for her driving but because she was a woman. So we were "unindicted coconspirators" in her gender campaign. It should also be pointed out that her Indy career began at a very advanced age. Most drivers who turn 45 are at the end of their racing careers, and Lyn was just starting her Indy career. She continued until she turned 53. That counts for something!

Her career prompted many snickers and back-room jokes poking fun at the fact that she was a woman (and many of those jokes included that she was a slow woman, to boot) and in some ways changed the landscape and environment for the next female driver, Sarah Fisher.

❖❖❖

ALL IN THE FAMILY

Before the race started in 1992, Al Unser Jr. wondered if he had the right chassis package. When we talked before the race, I could sense his doubts. But Al reminded me that the 500 was a race where you had to just be in a position to win and then let fate take over.

The race really belonged to Michael Andretti all day, but he lost fuel pressure, setting up a shootout between Little Al and Scott Goodyear. All you had to do was listen to the crowd to hear how close a battle it was on the racetrack. I call it the Indy wave. As the cars went into turn 1, you could hear the fans in turns 1 and 2 and on the short chute cheering wildly.

The front stretch stands gradually went silent. Then a hush fell over the turns 1 and 2 stands as both Al and Scott raced down the backstretch.

Then off in the distance you could hear the thousands of fans watching the race in turn 3 start to get loud. It was almost like hearing thunder in the distance.

Tom Carnegie, the track PA announcer, kept everyone up to date with short sentences.

Scott Goodyear's car (#15) was just shy of beating Al Unser Jr.'s. It was the closest margin of victory at Indy.
(PHOTO COURTESY OF THE INDIANAPOLIS MOTOR SPEEDWAY.)

"It's Unser!" he said in his baritone voice. "Will Al Unser Jr. win the Indianapolis 500 or will it be Canadian Scott Goodyear?"

Carnegie's call started to get drowned out by the buzz developing as the cars entered the short chute up in turns 3 and 4 for the final time. It got louder and louder as Unser's blue and white Valvoline car swung wide out of turn 4 and headed to the yard of bricks at the start/finish line. The noise became deafening as these two guys gobbled up track at the rate of a football field a second.

Then halfway down the front chute, Goodyear swung wide and made his move for the lead, a last attempt to claim glory. The crowd was going nuts!

They crossed the line.

But even though I was positioned in Victory Lane, which was directly across from the finish line at the foot of the old timing and scoring tower, I couldn't tell who had won when Goodyear and Little Al crossed the line!

They were just a blur, and I waited along with about 400,000 folks to see whom I would be interviewing.

"He's won it!" said an excited Tom Carnegie. "Al Unser Jr. has won the Indianapolis 500 in the closest 500 finish in history!"

I don't think Al expected to win or viewed it as his time. Al and I always talked about how fickle Indy could be and how he never thought he would win one.

So when Little Al rolled into Victory Lane, it was special to me. I knew him very intimately, and we had shared many of our stories about growing up around Indy. I knew what it meant to him, and he knew that I knew.

All I had to say was, "Welcome to Indy's Victory Lane."

He was in tears, and he said, "Jack, you just don't know what Indy means to me."

The tears kept flowing down his face, and all I said to him was, "Yeah, I do."

That win was incredibly important to Little Al because now he was officially part of the Unser racing clan.

❖❖❖

SAYING GOODBYE

A.J. Foyt shocked a lot of people when he hung up his helmet in 1993. He said goodbye to his beloved Indy the way he did so many things: quickly and on the spur of the moment.

When practice started for the 1993 race, A.J. was in his garage just to the right of the Gasoline Alley entrance and in the first row. His practice times were decent, but

Tom Carnegie interviews A.J. Foyt after his final lap in an Indy car at Indianapolis Motor Speedway. A.J. borrowed teammate Robby Gordon's sunglasses to try to hide the tears flowing down his face.
(PHOTO COURTESY OF THE INDIANAPOLIS MOTOR SPEEDWAY.)

they would not put him in contention for a front-row starting spot. Just two years before, he started in the middle of the front row. And in 1992, he finished ninth in the race—so there was still plenty of racing left in the guy.

But when dawn broke for the first day of qualifications, word spread that Foyt was going to officially retire and take his final lap around the track and say goodbye.

I rushed over to his garage, and there was A.J. still in his civvies. His crew was working on his car just like every other team in Gasoline Alley.

"A.J.," I asked, "is it true?"

He never answered me.

Instead, he walked over to the crew, barked out some orders, and went to the other side of the garage.

I'd been around Foyt enough to know that this was the end. I also knew that in about two hours I would be witnessing the end of an era. I left the garage and went about my business. Around 10 a.m., track announcer Tom Carnegie made the solemn announcement as Foyt's crew fired up A.J.'s car and it pulled slowly off pit road.

"Ladies and gentlemen, the legendary A.J. Foyt has decided to retire."

The crowd stopped what they were doing.

"Please congratulate him with your applause as he takes his final lap here at the Indianapolis Motor Speedway."

The crowd didn't applaud. They cheered! It was thunderous, and it drowned out the low-revving Chevy that sat beneath the cowling on Foyt's car as he drove it slowly around the track.

When Foyt got back to pit road, he spoke to the crowd and with tears in his eyes explained that the time had come for him to move on.

"I'm not retiring," he said. "I'm just not going to drive Indy cars anymore."

A.J. always wanted to leave the door open. Like many great athletes, he needed an escape door if he decided retirement didn't suit him.

Later in his garage, he admitted to me that while he was driving around the track that final time, doubts had surfaced.

"I thought to myself, what are you doing, A.J.?" he told me. "I almost said to hell with it and was gonna just mash it, but it was too late."

❖❖❖

A CLASS ACT

Roger Penske has always said, "Winning is where preparation and hard work intersect."

But preparation came back to haunt him at the 1994 Indy 500 a year after he had unveiled a special Mercedes pushrod engine. His drivers—Al Unser Jr., Emerson Fittipaldi, and Paul Tracy—dominated in 1993. Between them, Unser and Fittipaldi teamed to lead all but seven of the race's 200 laps, and Little Al won the race, his second time in Victory Lane at Indy.

There was no competition that year, Roger's Mercedes engine was just too powerful. It was legal according to the rule book, but most thought it violated the spirit of the rules. There were a lot of people who left the event very embittered against Penske.

The next year, stock-block engines were outlawed, and Penske's cars struggled.

Penske called off Fittipaldi's lap during qualifying, and had he let it go, Fittipaldi would have made the field. But Penske didn't feel the car was going to be fast enough to make the show. Unser never came that close.

Although the team emptied the garage, Penske didn't pack up and go home after the qualifying failure.

Instead, he stayed at Indy and kept Fittipaldi and Unser with him there, too. The drivers weren't thrilled at having to stick around. The last thing a competitor wants to do is watch other drivers race for a prize that was meant for them. But Al and Emerson put on brave faces for the public and did everything that was asked of them. They met with sponsors and mingled with Roger's guests and attended all of the parties and functions they were asked to.

The Penske crew tries to get Emerson Fittipaldi's car to qualify in 1995. Ultimately none of Penske's cars made the cut.
(PHOTO COURTESY OF THE INDIANAPOLIS MOTOR SPEEDWAY.)

I've always believed, though, that if Roger wasn't their boss they would have been as far away as possible from the 500 that race morning.

They still showed such dignity in defeat that I became a huge Roger Penske fan.

❖❖❖

CRASH COURSE

Indy cars are designed with what engineers call deformable structuring. In a crash, they are designed to shed their parts and pieces. The parts that come off the car take with them a lot of the energy generated by a crash and lessen the amount that is transmitted to the driver's area.

All of the cars that race these days carry a crash recorder like an airplane's black box that provides a record of the G-forces generated by a crash. Crashes at Indy can often record readings over 100 Gs.

But when Stan Fox crashed with Eddie Cheever Jr. in 1995, you didn't need a crash recorder to know that it was a bad one!

Cheever and Fox started 14th and 11th respectively that year. When the green flew, these two were mid-pack as the field funneled into turn 1. That's when all hell broke loose.

Fox collided with Cheever. Carlos Guerrero and Lyn St. James were also involved, but right from the beginning, the one you were terrified for was Fox. Stan was a diminutive journeyman driver who always seemed to be at Indy and always ready with a quip and a hearty handshake. He

Stan Fox and Eddie Cheever collide at turn 1.
(PHOTO COURTESY OF THE INDIANAPOLIS MOTOR SPEEDWAY.)

Fox's car is launched in the air as it disintegrates.
(PHOTO COURTESY OF THE INDIANAPOLIS MOTOR SPEEDWAY.)

Fox's car slams down on the track. Fox, who was seriously injured, survived the crash.
(PHOTO COURTESY OF THE INDIANAPOLIS MOTOR SPEEDWAY.)

would always ride around Gasoline Alley in a golf cart decked out to look like a bowling pin in honor of his sponsor being the Bowling Congress. His best finish was a seventh his rookie year in 1987, but that never kept Stan from his dream of Indy success.

In 1995 he was having a good year and had his best shot at Indy. He was driving for Ron Hemelgarn and was solid throughout practice and qualifications.

But none of that mattered in turn 1 of the first lap as his car disintegrated—piece by piece—first exposing his limp legs and then stripping away the carbon fiber until his lifeless body was almost fully exposed. He was unconscious and slumped forward.

Miraculously, Stan was still alive, and Indy's medical crew got him out of what was left of his car and transported him to Methodist Hospital in downtown Indianapolis.

He remained in a coma for some time but eventually woke up. He was never the same, though. Stan Fox never drove at Indy again, but until his death a couple of years ago, he was always in Gasoline Alley in May with his smile and ready for a handshake.

After his accident he co-founded an organization called Friends of the Fox, which raises funds for head injury victims. Stan, who was killed in a car crash in 2000, always hosted a group of head injury survivors every year at the track—giving them tours of the Indy garages and sharing with them his memories of the 500.

❖ ❖ ❖

AT HIS OWN PACE

Scott Goodyear qualified his LCI Honda on the outside of the front row, and all month long he was quick. You just knew that 1995 was going to be Scott's opportunity to get even for his split-second loss to Little Al a couple of years before.

Right from the start of the race, Goodyear's car was almost perfect. He didn't lead a lot, just hung around near the front. But you knew that his was one of the cars to beat.

Indy car drivers will tell you that the Indy 500 is run in segments. The first segment is qualifying. The second is the first 400 miles of the race, and the last and most important segment is the last 100 miles.

That was the way Scott raced that day. He stayed out of trouble, and when Scott Pruett crashed with 16 laps to go, it was Goodyear leading the pack behind the pace car.

When the race went back to green, the pace car just didn't look like it was up to speed. It was down on the apron, and Goodyear flew past it before the green light went on. Back in 1995, USAC was in charge of things at Indy, and they were a bit inconsistent in their rule enforcement. In some respects they suffered from the fact that the Indy 500 was the only Indy car race they officiated at every year. The rest of the Indy car schedule was run by CART, and their officials, I believe, had a more intimate understanding of the hows and whys of an Indy car race.

Well, USAC decided that Goodyear deliberately passed the pace car, so he was black-flagged.

Here was Goodyear, just cruising along, when a pace car driver denied him a chance at the biggest moment in his driving career!

Scott Goodyear and his wife, Leslie, cry after he is black-flagged for passing the pace car. (AP/WWP)

It was wrong, not in the letter of the rule, but in the intent. Goodyear was eventually dropped from scoring and ended up officially finishing 14th that day.

But in my heart, he had won.

Years later, the debate still rages on as to whether the pace car driver did not have the car up to the agreed-upon speed when Scott passed him. It sure looked that way to me, but USAC intervened and Goodyear ended his driving career without ever winning the Indy 500.

❖❖❖

THE MAGIC EYE

When it comes down to trying to figure out what will happen in a race, Bobby Unser is the most proactive.

He looks at pit stops and looks at the second series of stops at the Indy 500 and can come back and tell you what it means down the road for the finish of the race.

In 1995, Jacques Villeneuve was a lap down because of a pit violation early in the race that led to a penalty, and as soon as it was assessed, Barry Green, Villeneuve's crew chief, adjusted his strategy and began assuring his driver that all was not lost. Barry just stayed calm on the radio and kept telling Jacques not to worry, that there was plenty of time left in the race and they would be able to get the lap back.

Always analyzing, Bobby realized just by the lap times Jacques was turning that he wasn't out of the race. So Bobby predicted during the race that Jacques was going to get his lap back because it happened so early in the race.

Now there was a lot of stuff Bobby would say that we just laughed off as Bobby being Bobby-like.

"You've got to remember," he always explained like a teacher to a student, "when a driver pulls back out onto the track after a pit stop, it takes him at least a lap to get his car back up to full rpm and speed. Engines are like giant flywheels, they have to get mo-men-tum."

That's the way he'd say it—making momentum a long, drawn-out word as if to emphasize his point.

But very few people doubted Bobby when he said Jacques would come back and win the race. We went with Bobby's projection and started to follow Villeneuve closer than we normally would when a car was a lap down in the race.

And Bobby was right. Jacques came back and won.

Those insights were the things that made Bobby Unser both a great race car driver and a very, very good analyst on television.

❖❖❖

CAUGHT IN THE SPLIT

Tony George decided he wanted his own racing series in 1995, which led to the split of CART and the creation of the Indy Racing League.

BecauseTony's family owned the Indianapolis Motor Speedway, he held a card that the CART folks didn't understand the value of. They never expected Tony to be able to parlay the Indy 500 into a position of power where he could make sure that the Indy Racing League not only survived, but over a period of years flourished.

What distressed me when it all started was that people made it personal. You either had to be for CART or against CART: If you had any interest in the Indy Racing League, you were the dedicated enemy of CART. And if you had any interest in CART, you were the dedicated enemy of the Indy Racing League.

For the most part, that line in the sand was not drawn by competitors. It was drawn by people in the media and in affiliated companies who felt it was time to exercise their personal agendas and grew from there.

Tony made two mistakes when he first broke free from CART.

First, he wasn't really good at bringing in good people to birth his new league. Jerry Hauer was the first IRL director, but by the first IRL race, Jack Long had replaced him. There was no question that Jack had a personal agenda. But this was a time when he and Tony were putting together something new, and they needed to check all ambitions at the door.

I always got the feeling that Jack had locked horns with CART, and the IRL was an opportunity for him to come in and show them how to do it right.

Tony George wanted to protect his family's race in 1995.
(PHOTO COURTESY OF THE INDIANAPOLIS MOTOR SPEEDWAY.)

The second was when CART said, "We're going to consider boycotting the Indianapolis 500," the knee-jerk reaction from Tony and the IRL was to ensure there were 33 cars for the 500, by creating a rule that guaranteed the top 25 teams in the IRL would make the field. He was just trying to protect his race—not the series, his race. CART never really gave adequate consideration to what Tony wanted for Indy. They also felt that this, along with other IRL actions, was running counter to what they wanted, which was temporary street circuits, more road racing, and becoming an alternative to Formula One.

So it became this battle of words and ideas, tit for tat, back and forth to where the whole thing spiraled out of control.

So out of what I thought was a very, very difficult time, the IRL was born. And the drivers were caught in the crosshairs.

My dad always told me the only thing a driver has invested is his helmet, and he gets that for free. I always disagreed, because a driver has his career invested in it.

So what happened was the Michael Andrettis, the Al Unser Jrs., and the big stars had the Indianapolis 500 taken away from them in one big swoop. There was nothing they could do about it.

❖❖❖

INSIDE THE SPLIT

During the CART-IRL split, Al Unser Jr. was struggling with his marriage and battling alcoholism. He also was under contract with Roger Penske when Roger decided to support CART, and that meant not participating in the Indy 500.

The race, Al Jr.'s beloved Indianapolis 500, was the most important thing in the world to him. Now, through no action of his own, it was gone at the time when he was in his prime as a racer.

With Roger's choice, Al Jr. had to become one of the CART leaders in the fight for open-wheel supremacy. Now the race car was no longer the solace he sought. It became a burden. It was a downhill slide from there.

Al wasn't winning in CART. Penske wasn't winning either, but Al had to keep up appearances and answer a whole bunch of questions about the future of CART and the IRL. And he still wasn't winning.

There was too much politics and not enough pure racing to satisfy him. He wasn't at Indy, and yet he was with one of the top Indy car teams in the business. I have always compared Al's exile from Indy at that time to the enforced absence of Muhammad Ali in boxing while he fought the United States over his refusal to enter the draft.

Both lost some of their prime years of production to forces outside the ring and the racetrack. And it totally changed Al. He had always been the kid with the light in his eyes. I would run into him on the CART circuit, and it was gone.

The kid was gone.

❖❖❖

WALKING THE LINE

The hard part for me about the CART-IRL split was that ABC decided that the race reporters were going to cover both series.

Now I understand why, but my father taught me that if you walk down the middle of the road, you are going to get hit by trucks coming both ways. And, of course, that was what happened.

You would go to an IRL race, and there would be people who felt that you favored CART more with your commentary, with the way you spoke, and with whom you talked. Then you would go to a CART race, and there would be people who felt that you were completely for IRL. And no one participating in either group would believe otherwise.

It was the most difficult time in my broadcast career. No one would understand that I was there to do my job and I enjoyed both series.

❖❖❖

GETTING A GRIP

Although the sport had become "Penske-ized" through improved technology, engineers, aerodynamists, line mechanics and team managers, A.J. Foyt was still "Super Tex." He used his brain and experience to make the team competitive.

It didn't matter what kind of an argument someone had to try to convince him otherwise. For example, Goodyear was supplying tires to all of the teams, and they assigned an engineer who had studied dynamics and chemistry to A.J.'s team.

The engineer was trying to explain why the tires had lost their grip after practice, but A.J. refused to listen.

"Boy," he growled, "do you know who you are talking to?"

"A.J.," the engineer stammered, "I have an engineering degree, and I'm telling you..."

"I don't give a damn what you've got," A.J. interrupted, "I'm the one out there bustin' my ass, and I'm telling you what I want!"

That engineer backed out of A.J.'s garage, and I never saw him the rest of the month.

❖❖❖

RIVAL RESTART

Competing against the first Indy 500 after the split was the infamous U.S. 500, which ran the same day in Michigan. CART had this attitude that they were the best drivers in the world, and the IRL only had slugs.

And when you looked down that 1996 Indy 500 starting list, there were some names that made you wrinkle your brow a little bit.

Well, during the pace lap at Michigan, there was a huge crash—and a lot of the CART drivers had to go back to their garages.

That fact spread up and down pit road at the Indy 500 like wildfire. Crew members, on all of the teams, would get the news from a fan or someone back in Gasoline Alley watching the U.S. 500 on TV.

"Did you hear about that U.S. 500?"

"Big crash at the start! Half the field was wiped out! And we are supposed to be the amateurs."

Then the crew member would just shake his head in disbelief and go on to his job.

I first heard about it when Bob Goodrich, our ABC producer, told all of us during a commercial break.

"FYI, guys, the CART race had a big crash right at the start. They've red-flagged the race, and some of the drivers are going to have to go to backup cars."

I was surprised and, I'll admit, just a bit amused. Even the best in the business can have a bad day.

❖❖❖

PUSHING HIS SWITCH

No matter what you said, Bobby Unser always wanted to be the one to add to it. It got to be a joke with us in the TV business because Bobby always had to have the last word. So during the race about two years before he retired from announcing, we told him that Paul Page had a special switch upstairs in the main announcer booth that could turn Bobby's mike on and off.

It was the first time Bobby didn't have the last word, because we had him convinced that as soon as he was done talking, Paul was going to turn his microphone off and his additional commentary would not make it on air.

After the race he came down from turn 2 and was complaining about how we had to get rid of the switch because it was limiting what he wanted to do. Lo and behold, the switch never existed at all! But it worked like a charm.

We never told Bobby the truth until after he retired from the broadcast team. When he found out, he was at first a little pissed off. But then he realized what a good joke it was. He joined us in laughing about it. To this day, when we get together with Bobby, we always kid him by

Bobby Unser left his race car and went straight to the announcer's booth. (PHOTO COURTESY OF THE INDIANAPOLIS MOTOR SPEEDWAY.)

asking if he has checked with Paul and gotten Paul to turn on his switch.

❖❖❖

SO CLOSE, YET ...

Tony Stewart took Indy by storm in his first Indy 500, qualifying on the front row and competing for the

Tony Stewart focused on claiming Indy in the name of his teammate, Scott Brayton.
(PHOTO COURTESY OF THE INDIANAPOLIS MOTOR SPEEDWAY.)

victory. Tony and his Menards team were extremely motivated that year. Scott Brayton, the pole sitter and his teammate, was killed earlier in the month in a practice accident.

Scott had loved Indy. His family built racing engines, and he grew up going to the track. When his memorial service was held the week leading up to the race, his widow, Becky, spoke about Scott's love of the 500 and the speedway.

When John Menard offered to withdraw his entries in the race out of respect for Scott, Becky and Scott's dad said, "No. Scott would never have wanted that." Instead they suggested, "Just go out and win it for Scott."

I think Tony was just a bit torn about going on with the race. With Scott gone, the spotlight was thrust right

Tony Stewart leads the pack.
(PHOTO COURTESY OF THE INDIANAPOLIS MOTOR SPEEDWAY.)

onto Tony. Coupled with the Brayton family's wishes, the 1996 race was about Tony racing for Scott's memory.

But they also had to go on with the show. A lot of other teams, who were using older equipment, were scrambling for parts because 1996 was the year before the IRL instituted their own chassis and engine programs. Everybody knew the Menards teams had the best equipment. The Menards boys were running the Menards V6, and it was very fast.

Tony had led 44 laps and looked good to go on and take the win that everyone had hoped his team would grab. But, alas, the Indy gods had other plans.

There wasn't any warning. Just all of a sudden, Tony slowed and pulled into the pits.

The engine in his car had let go, like it did every year that it ran at Indy. It just didn't last for 500 miles.

His day was over, and Tony got out of the car and got ready for his interview. That's when he said what I thought expressed how so many feel after an Indy disappointment: "I feel like this place just shit on me."

❖❖❖

SAVING FACE

The 1996 Indy 500, the year after the split, had a huge cloud hanging over it. That year, the equipment was a year old, so the minute something blew up, crews were running around in the middle of the night trying to find replacement parts. There was a move to try to choke down the supplies because if the teams didn't have the equipment, they wouldn't be able to race.

There were also many rookies and unknowns in the field. These guys didn't know enough about Indy and racing there. It was like taking a bunch of kids playing stickball in the street and saying, "You are in the World Series. Go out there and win."

Everywhere you turned, people were saying that this version of the Indy 500 was an inferior product, and you began to wonder if it was true.

The only thing that saved the race that year was the win by Buddy Lazier. When Buddy arrived at Indy, he walked slowly with the aid of two canes because of a back injury. He could not stand erect for more than a couple of minutes before he had to sit down. He took daily therapy treatments for the injury, and his crew fashioned a special seat with additional lumbar support after consulting with a number of orthopedic experts. His back was broken like an egg that had been dropped off a wall. He was in such

Buddy Lazier's win in the face of physical pain paved the way for Indy and the IRL to continue.
(PHOTO COURTESY OF THE INDIANAPOLIS MOTOR SPEEDWAY.)

pain that he couldn't even sit up in the car. There was concern that Buddy wouldn't be able to race, let alone win.

Only five drivers led that race—and all five were veterans. Alessandro Zampedri ran the year before and led 20 laps. He, Buddy, Davy Jones, and Roberto Guerrero fought it out down to the wire.

These guys went at it all afternoon. There were some real spine-tingling moments—not all of them the kind you ever thought you'd see at Indy.

Late in the race while Davy was leading and trying to pass Eliseo Salazar, who was a lap down, the Chilean pinched Jones tight against the pit wall barrier. Jones didn't lift and saved the car and his race, but it brought a collective gasp from the crowd.

On the last lap of the race, Roberto Guerrero was in the running until he crashed in turn 4. Guerrero's crash brought out a caution, and Buddy swept under the checkered flag as the starter also waved the yellow, ending the race under caution.

It was such an important victory because Buddy was a journeyman driver whose dad had raced once at Indy. He didn't have a fat checkbook to bankroll his racing with and had spent a few years in CART driving outdated equipment while others came on the scene from Europe and South America and got better rides because they brought sponsorship dollars with them.

I asked Buddy a couple of years later why he decided to play roulette with his health and compete in that race, and he corrected me, saying, "There wasn't much more damage I could do to my back. It was just a question of how much pain could I endure. I'd take all the pain I felt driving that car and double it for another Indy win."

And his win was just enough of a cushion for the IRL to survive.

❖❖❖

ON THE CUTTING ROOM FLOOR

If there's one thing that has probably kept the Indianapolis 500 through the years in the lofty position that it holds, it has been the television coverage. But there are always little snafus.

One year, Bill Fleming was doing the pits and he went to Tyler Alexander, who at the time was Tom Sneva's crew chief. It was the first time that we had introduced in-car radios, and we were hearing the radio transmission between Tyler and Tom Sneva.

After listening in, Bill Fleming said, "We're with the crew chief Tyler Alexander, and Tyler, was that Tom Sneva you were talking to on the radio?"

Tyler replied, "No, you asshole, it was someone in the crowd. Who do you think it was?"

Fleming had a pretty big ego, and as soon as Tyler came at him with his sarcastic response, you could instantly see the shock on Fleming's face and the embarrassment over being chastised in such a vulgar way.

And Tyler knew exactly what he was doing when he made the remark. That was just Tyler Alexander's way. He was a contrarian and never cooperated with the media—especially TV.

Needless to say, that dialogue never aired. There are some benefits to editing.

❖❖❖

CAUGHT ON FILM

When Indy rebuilt Gasoline Alley, they redid part of the area behind the pitside grandstands, including a cafeteria where all of the crews used to eat lunch. Part of the renovation was adding some photo montages of the old garages, including random scenes of crews at work. They were just guys working in T-shirts and greasy hands. Nothing special for Indy.

Well, one day I looked up, and in a photo there was this big ol' tall guy, just covered in grease all the way up to his elbows. I thought nothing of it until I recognized him. It was a young Bill France Sr.!

It struck me as a bit ironic: The father of stock car racing was also attracted to the Indianapolis 500! I did get it confirmed that it was Bill Sr., but nobody knew when the picture was taken or what crew he had worked on.

Unfortunately, I discovered it after Bill Sr. had died and will always regret not having had a chance to ask him about his Indy memories.

❖❖❖

DON'T SHOOT THE MESSENGER

A.J. Foyt showed me an Oldsmobile engine from Jack Roush that they had just pulled out of the car. It had a hole through the right side of the block, and there was also a hole on the left side.

As soon as I had walked into A.J.'s garage, he pointed to the ruined part that lay on the floor.

"Ain't much left of that $75,000," he said.

"What happened?" I asked.

"Dunno, but the damn hole is big enough to see you from the other side!" (I later learned that the engine had blown up and a piston had poked through the block.)

Foyt slapped my shoulder and said, "Now that'd be a picture. Get your cameraman to interview you pointing the camera through them two holes."

I thought it would make a good visual, so I did it. It ran that day. After I got back to my hotel that night, the phone rang.

"Jack, this is Jack Roush," he said.

"Hello, Jack," I replied.

"You're killing me!" he complained.

"Excuse me?"

"I saw your report from Foyt's garage today. I can't believe you threw me under the bus like that ..."

"Wait a minute, Jack, I, uh," I stammered. But he cut me off.

"We have built a lot of engines for Indy this year, and the only ones that we seem to be having a problem with are the ones that are going into Foyt's cars. Come on Jack, you know how A.J. likes to tinker with things. Until I get that engine back here and have my guys go through it, I won't know what the problem is. But geeez, did ya have to do that this afternoon?"

A.J. had wanted to let Jack know that he wasn't happy with what Jack's engine-building company was sending him. And I had gotten myself caught in the middle!

❖❖❖

FAITHFUL FELINES

There is a group of fans that sits just outside the exit of Gasoline Alley on pit road.

They are there every day that the track is open, and they cheer and wish every driver well as he or she walks or drives out for practice, qualifying, or the race. They call themselves the Alley Cats, and the group is a fixture at Indy. They sit there wearing shorts and shirts honoring their favorite drivers, and they have all of the track programs and the daily newspaper so they can be fully updated on what is going on.

The Alley Cats on their stoop.
(PHOTO COURTESY OF THE INDIANAPOLIS MOTOR SPEEDWAY.)

When the 6 p.m. gun goes off every day, they offer up a small libation to their heroes. They call it having a capful for good luck. I don't drink, but every May I make an exception. I stop by and share a capful of Jack Daniels with the Alley Cats to celebrate another Indy 500.

You can always find them because they are in the same spot every day, never straying from their perch adjacent and to the direct left of the exit from Gasoline Alley.

But when the new Formula One garages were built, the speedway security staff tried to relocate the Alley Cats from their territory. The yellow shirts wanted to create a more open area and moved the Alley Cats back about 25 feet to the edge of the new F-1 ground level suites that abutted Gasoline Alley.

On the first day the track was open, Mari Hulman George drove by in her golf cart and saw the Alley Cats in their new location.

"Hey Mari!" they shouted. "Great to be back; keep up the good work! We're with ya!"

Well, they caught her attention, so Mari stopped and asked why they were at the edge of the building instead of in their usual spot.

"Yellow shirts say we can't be there no more!" they shrugged matter-of-factly. After chatting with them, Mari left a couple of minutes later. The next morning, when the track opened, there was a brand new barricade set up where the Alley Cats had always stood with fresh new paint that read, "RESERVED FOR THE ALLEY CATS."

They've been there ever since.

OUTPOURING OF SUPPORT

A rookie named Greg Ray showed up at the Indianapolis Motor Speedway in 1997 with Thomas Knapp. They had cobbled together an entry and set about the task of trying to put Greg into the field.

No one knew much about Ray, other than his family owned a very successful marine sales company in Plano, Texas. And Thomas and Greg had no sponsorship for their car. That, coupled with their very, very fast practice speeds in the early sessions, drew a lot of attention to them, and I set out to learn more about the rookie.

Five years earlier, Ray had started his career by winning the 1992 SCCA Formula Amateur Ford 2000 Championship. He climbed quickly through the ranks of road racing and set out to fulfill a dream that had formed late at night back in 1990.

"I'd sit up at night and watch ESPN," Greg explained. "Every night, there was a series called Legends of the Brickyard. That's where I first discovered Indy and fell in love with it. I wasn't sure how to get there; I just knew that I wanted to get there."

So Greg enrolled in a driving school and like all of the other students got a chance to state his reasons to the rest of the class as to why he was taking the course.

"Hello. My name is Greg Ray, and I'm taking this course because I'm going to race and win the Indianapolis 500!"

Of course, there were a lot of snickers from his fellow classmates, but that all ended by 1997. Ray was at Indy and ready to make his first start.

His story got a big boost that season because of rain. The wet weather meant no on-track action, and our TV cameras went searching for compelling stories. Ray's was one of them. We repeatedly covered his quest and highlighted his lack of funds. After the first day that we aired the story, fans rallied to support Ray.

Several days later, Greg called me into his garage.

"Jack, take a look at this," he said as he pointed to a stack of mail delivered earlier that day. "People are sending money, checks, cash, you name it!"

Greg was overwhelmed by the display of generosity, and when he pulled his car into line to qualify, he and Knapp taped two $100 bills to the side pod area normally reserved for a car sponsor's logo.

After qualifying, I searched out Greg to finish the story and fan angle. I couldn't find him in his garage and asked Knapp where he was. Knapp pointed up to a small storage area above the workbenches that bordered the front wall of the garage stall.

Tucked up there, amid the car covers and spare wings, was Greg Ray—sound asleep!

❖❖❖

"GENTLEMEN, START YOUR ENGINES ... AGAIN"

In 1997, the funniest thing I ever witnessed took place just before the start of the Indianapolis 500.

The command "Gentlemen, start your engines," was first used at the Indianapolis 500. Today, you hear it at almost every race—big and small—but it never has the same solemnity that it has when it calls the 33-car field for an Indy 500 to fire up.

In 1997, Mari Hulman George, daughter of Tony Hulman and mother of Tony George, was still doing the honors. The race was washed out on Sunday and rescheduled for Monday, May 26. Like everything in Indy's pre-race, the command is choreographed and timed down to the second.

Until his death in late 2003, Jim Phillippe was always the public address official in charge of introducing who was going to give the command.

That year, like every year, a track official gives the signal only after race control has confirmed that every driver is correctly strapped into his or her car and ready. That year it was USAC's Mike Devin.

Mike had done this for many years without a hitch. On that Monday, he was crouched to the right of the bank of microphones in front of Mrs. George.

Phillippe looked to Devin and whispered, "Are you ready?"

Devin hit the key on his radio to check with race control, and Phillippe thought he was signaling that all was ready for Mrs. George and started into his brief introduction.

"And now race fans, for those famous words you have been waiting for..."

As Phillippe was saying this, a look of pure terror came over Devin's face. He started waving and shaking his head and trying to get Phillippe's attention.

"No! No! Jim, we aren't ready!"

But Devin didn't shout his instructions. Perhaps out of reverence for the moment, he whispered them!

So Phillippe continued with his spiel.

"Here is the chairman of the Indianapolis Motor Speedway, Mari George!"

Mrs. George didn't miss a beat and responded, "Gentlemen, start your engines!"

At that point, Devin just threw up his arms in surrender and just laid his head on the floor.

Part of the field rumbled to life while others just sat there with their engines off.

A couple of seconds, and all of the engines were silenced and everyone—including Mrs. George and Mike Devin—waited for the signal. Less than a minute later, race control gave the signal and Mrs. George called the cars to life again!

It set the stage for a wacky day. Only 16 laps were run before rain washed out the rest of the day, and the final 184 laps were run on Tuesday.

Arie Luyendyk won his second Indy 500 that Tuesday after three commands—two on Monday and an additional one on Tuesday—to "start your engines!"

❖❖❖

WATERED DOWN

In 1997 Affonso Giaffone's team was refueling while I was covering their stop in the pits, and the fuel nozzle malfunctioned. When the crew member who manned the fueling hose removed the nozzle and hose from the car, fuel sprayed everywhere and doused my pit uniform. I was soaked from my neck to my waist in fuel and surrounded by methanol-burning Indy cars.

It all happened so fast that no one had any time to even say anything. The crews were working, I was calling a pit stop, and it was just business as usual, except I was covered in fuel!

But I was very worried that it might ignite. One of the crew members just threw a bucket of water on me to dilute the fuel.

I didn't mind getting wet. I think I even said, "Thanks, man! I needed that!"

❖❖❖

COMPUTER CRASH

In 1998, A.J. Foyt's driver, Kenny Brack, qualified on the outside of front row and at the drop of the green flag raced to the front.

Brack was burning up the track, and it was looking like it could be the Swede's day.

Tommy Lamance, A.J.'s nephew, was in charge of calculating fuel mileage. Now, most Indy car teams have a cadre of engineers to extrapolate all of the telemetry data that an Indy car generates. In real time, they can see what the engine revs are. They know the tire pressures, the oil pressure, and more than 250 measurements in the blink of a megabyte.

It was almost time for Brack to pit, and A.J. didn't need a computer to tell him that he needed to bring his driver in for fuel. He has a sixth sense about these types of things. But Tommy overruled A.J., telling his uncle that the computer from Compaq, one of A.J.'s sponsors that year, said that the car still had a couple of gallons left in the tank.

As Brack continued to lead the race, Lamance and Foyt jawed back and forth, debating whether to call him in. They yelled at one another until the radio went off.

Brack said, "I'm out of fuel."

A.J. exploded!

By the time Kenny limped onto pit road for his fuel, he was two laps down.

As soon as Brack got refueled and back onto the track, A.J. stormed over to Tommy Lamance's laptop computer and smashed it to smithereens.

Brack went on to finish sixth, but it was the faulty computer reading that cost him the race.

❖❖❖

TASTE OF THE RACES

It was Greg Ray's first Indy, and he was ready for a taste of speed and excitement.

But he wasn't ready for what happened next.

"They do, 'Gentlemen, start your engines,' and you come rumbling down, and if you're in the back row, the next thing you know after the first lap you're overcome by the fumes from the methanol being burned off."

Methanol makes your eyes water and your throat burn and it stings your nose. Seasoned veterans know to avoid inhaling it, but Ray didn't. And it gets kicked up in the air and stirred around by the cars running at over 200 mph.

Because the cars are open-cockpit, the stuff that you would normally see stuck to a NASCAR windshield over the course of a race has nowhere to go at Indy. That grime and film that gets kicked up all ends up hitting the driver.

So by the time the race was finally over, there was more than a pound and a half of debris on the car from all of the junk in the air.

"I don't think we ever got it all out until we finally retired the car," Ray chuckled.

❖❖❖

WITH OPEN ARMS

When the CART guys started to trickle back to Indy after the split, the only group that was upset was the media. The beat writers all sat there like they had been betrayed.

But the fans, the drivers, and the crews were happy to have even the most anti-Indy faction back on the track.

I'll never forget the return of Jimmy Vasser, the guy who stood in Victory Lane at the 1996 U.S. 500 and said, "Who needs milk?"

I fully expected people to give Jimmy a hard time about it. I mean, he was so brash at the U.S. 500 when he mocked Indy with his "Who needs milk?" statement. Even Jimmy thought things might be bad for him when he got back. He tried to defuse the situation in advance at the press conference announcing he was coming back. He sort of played on the original remark by saying, "I now need milk!" And it worked, because no one ever said a word and he was welcomed back.

There was never any animosity, and nothing negative happened to him whatsoever. He was simply seen as a driver coming back to Indy, and that seemed to be good enough for everyone.

Bygones were bygones.

❖❖❖

A WINNING SWEET TOOTH

In the spring of 1997, Jimmy Kite could only dream of the Indy 500. By 1998, though, he was in the Indy 500.

The way he got there is a story of perseverance. In 1997, Kite scored an upset win at Phoenix's Copper World Classic for USAC's Silver Crown Series by passing the race's leader, Chuck Gurney, in the final turn of the last lap.

He was so excited about winning that he parked his car on pit road because he didn't know where Victory Lane was. That win led to a meeting with Andy Evans, who was running the IRL, and Evans gave Kite a try at Pikes Peak later that year and a ride for the Indy 500 in 1998.

"I'd always dreamed about driving at Indy," Kite told me, "but I never thought I'd get the chance."

At about five feet, two inches, Kite looks more like a jockey than a driver of a 220-mph Indy car. His size, coupled with a great sense of humor, immediately endeared him to many veterans at Indy, but it was his practice crashes that made him a TV star.

Early in the month, Kite crashed his Royal Purple mount. He was so upset about the crash. He thought about how long his crew had worked on the car, and how he had given them more work to do in a shorter period of time so that he could get out on the track and practice.

"They worked all night to repair the car so that I could get back on the track and keep logging laps," says Kite.

Kite wanted to say thanks to the team for their hard work and came up with a sweet thank you.

"On the way to the track the next day, I stopped at a doughnut shop and bought four dozen doughnuts for the crew."

I told the story on one of ESPN's practice telecasts later that day while interviewing Kite and some of his crew members.

Well, a couple of days later, Kite crashed again. The next morning there were more doughnuts for the team. A couple of days later—you guessed it, another brush with the wall and more doughnuts.

Jimmy made the race that year, starting 26th and soldiering home to a very respectable 11th at the finish. That 11th spot was his biggest payday in his career. He earned $287,300.

"It's a good thing," laughed Kite. "I had a pretty big bill at the doughnut shop!"

❖❖❖

HERO WORSHIP

I was working in the press room doing the daily briefings the entire race month of 1999 because I had gone with the IRL to Fox, and the only event Fox didn't

Kenny Brack and A.J. Foyt bask in Victory Lane glory.
(AP/WWP)

cover that year was the Indianapolis 500. ABC had the race, and I was out in the cold.

I was devastated. I was dealing with the possibility that I had worked my final Indy 500 telecast.

Then lo and behold, it's the first race A.J. Foyt wins that I am there in person to witness.

A.J., as the winning car's owner, and Kenny Brack, his driver, were in Victory Lane as I struggled to watch it on a TV monitor, away from the action. He came into the press room for his media briefing, and it just wasn't the same for me.

Just as it came to a close, A.J. leaned in and whispered, "The one thing that was missing today was you in Victory Lane."

That meant a great deal to me.

❖❖❖

THE REAL KING

After his Indy win, Kenny Brack had finished his interviews, and I could see that A.J. Foyt was really worn out—the day had taken its toll on him. He looked really old as he waited for Kenny to finish. I could tell the pain in his feet from a crash he had years before at Elkhart Lake, coupled with standing on pit road all day, was taking a toll on him.

So he was sitting on his golf cart with his feet propped up. I could see him getting really frustrated waiting for Kenny. Just as Kenny was about to go to the golf cart and ride back to the garage with A.J., Kenny got a call on his cell phone.

It was the King of Sweden calling to congratulate Kenny on his win. I stuck my head outside of the door to tell A.J. "Kenny's going to be just another minute, A.J. He's on the phone with the King of Sweden!"

And A.J. snarled back, "I don't care who he's on the phone with. Tell him the King of friggin' Texas is tired and wants to go back to the garage!"

❖❖❖

FLAG MAN

Every year, there's a guy who parades around in a homemade outfit that is all checkered. He wears a toy helmet and goggles and parties around the infield—much to the enjoyment of all of the fans. He mugs for the fans, poses for pictures, and watches practice. He doesn't even have any tricks to support his act. Just his costume, a human covered from head to toe in checkered flag.

I've talked to him a couple of times, and he is just a typical fan—just happy as hell that it's May and he's at Indy. If you saw this guy walking down the street dressed normally, there would be nothing distinguishable about him. He's just a run-of-the-mill guy who loves checkered flags ... and Indy.

Still, he's really goofy-looking, but no matter how many times you see him, he's always good for a quick laugh.

❖❖❖

WHAT GOOD IS AN EMPTY CAR?

Billy Boat couldn't get his car up to speed on Bubble Day in 2000 and was in danger of not making the field for Indy. He had driven for A.J. Foyt and won the pole in his car in 1998, but A.J. had fired him at the start of the year.

But apparently the driver was still on A.J.'s mind. I was hanging out at A.J.'s garage, and he asked me if I thought Billy was going to make the race.

Indy's famous fan—the flag man.
(PHOTO COURTESY OF THE INDIANAPOLIS MOTOR SPEEDWAY.)

"I don't think so, A.J.," I said. "He just doesn't have the speed."

A.J. looked over at a spare car in his garage and said, "Have Billy come by the garage."

I delivered the message to Boat, and the next thing I knew, he was being fitted in the Foyt backup car. With less than an hour remaining in qualifying, A.J. rolled the car out onto pit road, had Billy run a couple of laps, and then told the crew, "Put it in line!"

They did, and Boat proceeded to put up a speed of 218.872 mph—good enough for 31st on the grid and a spot in the race.

He didn't win the race. But Billy finished 15th and earned $211,000. Not bad money for a guy who wouldn't have even been in the race if not for A.J.

❖❖❖

CLIFFHANGING

Doug Didero, an outstanding supermodified driver, came up to try Indy in 2000. His practice times were fast enough to make the field, but he and his crew wanted more. They were not satisfied with just making it; they wanted to be sure that they made it!

Then without warning, they fell off the speed cliff. They could never get the velocity back. They went out every day, and they tried and tried. But they couldn't recapture the speed they originally had.

And they never made the show.

❖❖❖

A CURSE OF A JOKE

In 2000, Mario Andretti was invited back to Indianapolis, during one of the qualifying weeks to make a ceremonial lap as part of the track's Legends of the Brickyard program.

The idea was to give people a look at some of the racing heritage and celebrate its long-standing traditions. Drivers and the cars that they made famous are reunited and take a lap around the track to the applause of today's fans.

Well, Mario was invited back to drive the car in which he won the race's pole position in 1966. I wasn't paying much attention to the ceremonial lap because I was busy taping some stuff for an ESPN telecast later in the day.

All of a sudden I heard, "Andretti is slowing down the backstretch!"

I first thought maybe it was a joke, because there was a long-standing joke told by many Indy folks that deals with Mario's bad luck and his Indy curse. It goes:

"What are the four most famous words in Indy 500 history?"

"Mario is slowing down!"

Mario also sided with CART in the CART-IRL split, and maybe the phrase was a jab for him being one of CART's staunchest defenders—he became a villain to many who took up for the IRL.

So I ran out to pit road to see if Andretti really was slowing down the backstretch, and sure enough, the car was stopped on the backstretch! I was dumbfounded.

I thought it was the ultimate insult to Mario. Fate could not even let him come back to the track and enjoy a ceremonial lap.

I thought about our 1985 conversation and said to myself, "Mario, there is an Indy curse!"

CART COMEBACK

When CART driver Juan Montoya won the Indy 500 in 2000, it was a great race. He and Buddy Lazier battled back and forth right down to the end, a classic Indy 500.

But no sooner did he win than you knew the CART crew was going to trumpet and herald the win as the real race car drivers vanquishing the cheap imitations that run in the IRL. See, because CART drivers stayed away for so long and the race went on with IRL guys who were considered second-rate talent, it just gave them ammunition when a CART driver showed up and won the event.

But that wasn't the case. No one was that upset that a CART driver won, and there was very little gloating. People were happy, and you saw in Victory Lane a guy who was really reserved, proud of what he had accomplished.

There were still people who wanted to make it a victory for CART versus the evil empire of the IRL. But for the most part, it was just a good victory for Montoya.

NEED A RIDE?

Donnie Beechler didn't have a ride for Indy in 2001 and was cruising around the garage with his helmet and firesuit looking for a job. Now, A.J. Foyt really liked Donnie—he had come up through the ranks driving sprint cars and silver crown cars. He cut his teeth on the dirt tracks of the Midwest. In 1999, he was leading an IRL race at Texas Motor Speedway when the pace car hit his car under caution, knocking him (and the pace car) out of the race!

So on a whim, A.J. decided to put Donnie Beechler in one of his cars and then in the show with a final day qualification speed good enough for the 27th spot on the grid. The money needed for Beechler's effort came right out of A.J.'s pocket.

❖❖❖

GEARED UP?

In 2001, a lot of the teams were having trouble with the forks that moved the gears in and out going from first to second, but no one knew what the fix should be. There were 16 units still out there in the pipeline, but they hadn't been tested, and so no one knew if they would work. It was like Russian roulette; you hoped you got a good one.

Well, Roger Penske bought all 16, X-rayed them, had his team—Gil de Ferran and Helio Castroneves—pick out the two best, and put them in the cars for race day.

Mark Dismore, who was with Kelley Racing, nearly dominated the race. When Mark came into the pits for his first stop, everything went on without incident. He got tires, fuel, and then the signal, "Go! Go! Go!"

But as Mark pulled out, you could hear that he was having problems keeping the car in gear. It chugged along and finally died. His crew had to push him back to his pit stall. Greg Ray took the lead, but Helio caught him and went on to win.

So once again, it was Roger Penske's preparation that got his driver into Victory Lane. By being proactive with the gearboxes, he found ones that would last an entire race and got Helio his victory.

❖❖❖

HAPPY HELIO

Helio Castroneves grew up racing karts with Tony Kanaan in Brazil. One year, he was watching the Indianapolis 500 when Emerson Fittipaldi won and the broadcaster said that Emerson had just earned $1 million.

Helio turned to his dad and said, "Look, Dad, they are going to pay him $1 million!"

I think it still blows him away when he cashes those checks for $1 million.

Helio first won the Indianapolis 500 in 2001, and I caught up with him five hours after the race in the Firestone tent. He was so genuine, had not taken his firesuit off, and was bouncing around.

"This is the greatest thing that has ever happened to me!" he said.

Helio Castroneves toasts his second Indy win the Indy way.
(PHOTO COURTESY OF THE INDIANAPOLIS MOTOR SPEEDWAY.)

After his victory, Helio Castroneves climbs the fence in celebration. His crew joined him shortly thereafter.
(PHOTO COURTESY OF THE INDIANAPOLIS MOTOR SPEEDWAY.)

Fast forward to his next win the next year, and his excitement was the same. He climbed the fence to celebrate. It never got old for Helio. He won the pole in 2003, and the guy who has already won two Indy 500s was overcome by emotion. For winning a pole!

This guy goes into every race feeling like this could be his last race because maybe it's all going to go away on him. And so he loves every minute of it.

❖❖❖

COMING HOME

Al Unser Jr. made his way back to IRL in 2000, but his results behind the wheel were not what they had been before the split. At Indy, he drove to try to taste victory again but never came close. He was obsessed, but he only placed 29th, 30th, and 12th. He began to wonder if he had lost his touch. He turned to alcohol. In 2002, he went to rehab, and when he got out, I was invited to the house of Tom Kelley, his car owner, in Ft. Wayne to interview him.

I landed at the airport, and Al Jr. was waiting in the car for me.

"Where's my $20?" he asked.

"Al, that's your friggin' father, not you!" I snapped back.

"Yeah, but I could use the $20," he retorted and smiled.

I saw the dance in his eyes again, and I knew then that the kid had returned. At dinner we talked for four hours—some of it on camera and some of it off camera.

Now he is just thankful to be alive and sober, and he is going to enjoy whatever number of days he has left in the cockpit. That doesn't mean he has lost his drive. Come race day, he wants to win just as bad as anybody else, but it isn't that obsessive-compulsive need that he had fallen

Little Al in the shadow of Indy in 2000 after a five-year absence.
(PHOTO COURTESY OF THE INDIANAPOLIS MOTOR SPEEDWAY.)

into when he came back to the IRL. He wants to enjoy every single moment of it and race as hard as he can.

After all, the season isn't over just because the checkered flag drops at Indy.

❖❖❖

MISSION TO WIN

Sarah Fisher was the first woman driver to drive at Indy who came up through the midget and sprint car ranks instead of road racing. When she was five, a local Ohio TV station did a feature on her quarter-midget driving skills and her races against boys.

In that feature, a gap-toothed, little blond girl said that all she wanted to do was win races.

Boy, did she win races.

Campaigning cars built by her dad, Sarah became a winner on the Midwest's short tracks and at age 19 got the call to test an Indy car. She became the youngest person to compete in IRL IndyCar Series history by racing in the season finale at Texas, starting 17th and finishing 25th after mechanical problems ended her race after just 66 laps.

Unlike Guthrie and St. James, Sarah purposely avoided the gender issue.

"I don't look at it that way," she has told me repeatedly since our first meeting. "The car doesn't know if it's being driven by a man or a woman."

In 2000, Sarah arrived at Indy behind the wheel of Derrick Walker's Cummins Diesel special and quietly went about the business of qualifying for her first Indy 500.

Lyn St. James was also trying to qualify for that race, so Sarah didn't have to have the camera solely focused on

Sarah Fisher races the only way she knows how—to win.
(PHOTO COURTESY OF THE INDIANAPOLIS MOTOR SPEEDWAY.)

her as the only female driver at Indy. Lyn was always eager to handle that issue.

Sarah's 220.237 mph clocking put her 19th on the grid. Back in her garage after that run, I caught up to her.

"Sarah, you've made your first Indy 500 and you haven't even turned 21 yet. What does a teenager do tonight to celebrate?" I asked her.

"I get to go home to my apartment and do laundry," she said.

"No celebrating?"

"Nope. It's about finishing my first Indy 500."

That accomplishment would have to wait. On race day Sarah got caught in a lap 69 accident with Greg Ray and ... Lyn St. James!

She got to finish her first Indy 500 a couple of years later when she drove her Team Allegra/Dreyer &

Reinbold Racing mount to a 24th-place finish in the 2002 race. After that race I spoke with her.

"Congratulations on finishing," I said.

"Yep, that means I'm just another loser," she said with a chuckle.

See, Sarah Fisher wants to win the Indy 500. She has no doubt—nor do I—that given the right equipment and right set of circumstances she will win the Indy 500! She is young enough to pay her dues and be ready for the opportunity when it presents itself. She still struggles to land full-time rides in top-notch equipment. One reason is that she does not play her gender up. She looks at her efforts and sells herself the same way a man does.

Sarah has won the hearts of the Indy car community. The seeds for that acceptance came in the 2001 IRL season opener at Homestead-Miami, where her late-race charge and pass of Eliseo Salazar in A.J. Foyt's car earned her a career-best finish of second.

Her pass on Eliseo in that race prompted Foyt to radio his driver one of the most famous transmissions ever. When Sarah put the move on Eliseo and took over second, an angry Foyt chastised his driver for letting Fisher get the best of him.

"You just got passed by a girl!" A.J. hissed.

For three consecutive years, she has been voted the league's most popular driver. In 2002, she became the first woman in North American motorsports history to win the pole position for a major league open-wheel race, capturing the MBNA Pole for the Belterra Casino Indy 300 at Kentucky Speedway. She led the first 26 laps and finished eighth.

❖❖❖

RACING PHILOSOPHY 101

A.J. Foyt has a theory on race car drivers. He told me most guys who drive a race car are just drivers, and every once in a while, a guy will come along who is a racer.

What he meant by that was you could acquire the skills to be an accomplished race car driver, but if you were a racer, you take those skills and put them on top of some God-given talent and become a Rick Mears, an Al Unser, or an A.J. Foyt.

He leaned over and said for the past few years, he had had drivers. Then, he pointed at his 19-year-old grandson and said, "That kid there, Jackie, is a racer."

A.J. Foyt (left) and Anthony discuss tactics and strategy during a practice session. (AP/WWP)

❖❖❖

MOVES OF HIS OWN

Anthony Foyt struggled as a rookie at Indy in 2003. His early Indy car efforts at Miami, Phoenix, and Japan were less than stellar, and not all of it was his fault. But Anthony took to Indy like a duck to water. Like his granddad, he added his own touch to his first drive there.

During a qualifying attempt, Anthony spun his car—not so unusual for rookies or veterans.

But what happened next was something that I had never seen before.

When his car whipped around 180 degrees exiting the turn, Anthony calmly kept it off the wall and proceeded to scrub off speed from 200 mph to 50 mph—running all the way down the backstretch backward!

The kid was unfazed by it all. He just got out of the car, and after a quick check at the infield care center walked back to his granddad's garage. I caught up to him just before he got to the garage.

"Anthony," I said. "What happened?"

"I dunno," he deadpanned. "I hit a bump, and bam, the next thing I know I'm going backward."

"Has A.J. said anything to you about what happened?"

"Not yet," he said. "But Jackie, you know Pa-Paw is gonna be pissed!"

Surprisingly, A.J. wasn't at all upset. In fact, I think he was a bit proud of the way Anthony handled the situation.

❖❖❖

PRESSED OFF!

The day after Anthony Foyt's faux pas, a couple of reporters tried to use the mistake to make the case that Anthony didn't belong at Indy. One writer cited a number of anonymous drivers who said Anthony was a danger to himself and others.

"Can you believe that crap?" growled A.J. Foyt. "It's a damn good thing that they didn't give their names 'cause I'd beat the crap out of each and every one of them!"

The hype gave way to the facts. Anthony qualified 23rd for his rookie Indy 500 and finished 18th, just two spots behind where his grandfather finished his rookie Indy in 1958. He stayed out of trouble all day long.

❖❖❖

NEED FOR SPEED

Mark Dismore grew up in Greenfield, Indiana, and came up through Karting. He dreamed of Indy but was not financially strong enough to get there. When he finally arrived, he was in old and outdated equipment.

Right off the trailer, Dismore's team was quick. But he wanted to show the world that he deserved to be part of the Indianapolis 500. That desire almost cost him his life. During the practice runs, I was on pit road, gathering bits of information for race day and listening to Tom Carnegie, the voice of the speedway for 59 years, shout out the individual speeds.

"Dismore! Mark Dismore is quick," he said and I figured that Dismore was catching a tow off other cars, an old trick that drivers use to get a fast time and hopefully some press.

Coming out of turn 4 his car stepped out and started to spin, hitting the wall and careening all of the way down the front stretch, flipping and disintegrating.

"We have a yellow!" Carnegie announced. "A car has crashed on the front stretch! It's Mark Dismore."

Then he went silent. I was about 300 yards down pit road, and I saw debris pile up at the entrance to pit road. Then Mark's engine—separated from the chassis—came sliding right by me! I watched as Dismore lay unconscious and was strapped in what was left of his car, a small capsule that left his legs exposed. He was half dead.

Years later, he told me why he ran so hard. "Because I wanted to show every owner who refused me a ride that they had made a mistake—a big mistake—in not hiring me. Instead, it was me who made the big mistake, and I almost died because of it."

❖❖❖

STROKE OF LUCK

Indy in 2002 went down to the wire. Helio Castroneves was leading when Paul Tracy passed him. Then they both noticed the caution flag had come out.

But when had the yellow flag been waved? If it had been before the pass, Castroneves would be the winner for the second year in a row. If after, Tracy would go on to Victory Lane.

Castroneves was declared the winner while the judges reviewed the pass. Timing, scoring, and a lengthy investigation showed Castroneves was still the leader when the yellow came out and his victory held.

It was a pure stroke of luck that the flag appeared. There is no question in my mind that Tracy would have won that race if it had gone green. Even if the caution had come out five seconds later, Tracy would have had it.

❖❖❖

EYE ON TALENT

Rick Mears is like a really good baseball scout; that's why his new role at Penske is so perfect for him. He spots for Roger Penske, and he researches whom Roger should hire as his next drivers.

I was on an airplane with him earlier in 2003 talking about Sam Hornish Jr., who was likely moving to a new team at the end of the season.

"It amazes me how adept Sam is at working the high side of race tracks when he is passing," I said.

Rick just smiled. "That doesn't surprise me."

Rick then went into graphic detail about how Sam uses split-second timing to get an extra drafting boost by precisely positioning his front wings right where they would get a split-second pull from the wake of the car he was following (and trying to pass) just before that car turned into the corner.

Rick Mears keeps his eyes open for new talent.
(PHOTO COURTESY OF THE INDIANAPOLIS MOTOR SPEEDWAY.)

The smile on Rick's face was huge. It was admiration, like a teacher watching one of his students excel.

I could tell he was studying Sam Hornish Jr., and I instantly figured out that if Sam left Panther Racing, he was headed to Penske to work with Rick.

❖❖❖

WALK OF FAME

One of the great traditions that Indy has is the walk from Gasoline Alley out to pit road.

Helio Castroneves takes the same walk that Jimmy Clark did in the 1960s. It hasn't changed except for the trappings.

Some things, though, remain constant. Whenever a driver starts out for pit road, Indy's yellow-shirted security force starts blowing their police whistles. The mulling crowd crossing the entrance to Gasoline Alley parts like the red sea—partly out of courtesy and in response to the whistles, but also out of anticipation that they will get a chance to see one of their heroes striding out to drive.

Programs, pieces of paper, and other things get shoved into the path of the drivers by fans hoping for an autograph. Others just shout out words of encouragement.

"Go get 'em, Al!"

"Arie, show them how it's done!"

"You go, Sarah!"

On race morning, this walk reminds me of the walk boxers make to the ring before a championship bout. The walk and the fan interaction is not just reserved for drivers.

A couple of years ago, I was walking out to the pits when fans started chanting my name.

"Jack A-Root! Jack A-Root!"

I'd be lying if I said that it didn't give me a bit of a rush. Just as I made the turn out onto pit road, a teddy bear on a string was lowered directly in front of me with a note that said, "Can you please autograph this for me?"

Here was this teddy bear, decked out in a blue blazer with a miniature ABC microphone, and a tiny headset.

After I signed the bear, I found the fan at the other end of the string. She explained that she was a big fan of the Indy 500 and had followed my broadcast career there.

"You're so good at what you do," she gushed. "I wanted to make this bear as a tribute to you. I call him Jack 'The Bear' Arute."

Now I ask you, how many guys have their very own teddy bear?

❖❖❖

GETTING HIS CAKE AND EATING IT TOO

Rain is Indy racing's worst enemy. Nothing happens on the track, but that doesn't necessarily mean all is quiet back in the garage.

That was the case in 2003 when rain washed out the first day of qualifying for the Indy 500. The ABC crew still had to put a show on the air, but without any racing action, we had to manufacture things to pass the time.

We set up a small studio in Gasoline Alley and shuttled drivers in and out, mixing interviews and footage from previous qualifying shows and features we had on many of the drivers. One of the topics we discussed was the emergence of Brazilian drivers, and we invited Tony Kanaan, Gil de Ferran, and Helio Castroneves to what we were calling "the rain room." We asked them to talk about how Brazilian drivers discovered Indy and why they were all choosing to race in the series.

But interviews only go so far, so I decided to spice it up earlier in the day when I set it up. Helio's public relations representative, Lisa Boggs, had given me the heads up that it was his birthday. I passed it on to a production assistant and sent her out to buy a cake that we could give him on the air.

With the cake off to the side, Gary Gerould and I greeted de Ferran, Castroneves, and Kanaan. We positioned them on the set, and Kanaan noticed the cake and whispered to me, "Jack, what is that?"

"It's a birthday cake for Helio. We are going to surprise him with it, so don't ruin it!'"

The look of mischief that instantly lit up Tony's eyes told me it was a lost cause!

Tony and Helio started their racing careers together Karting in Brazil. They moved up through the ranks together and are the closest of friends.

"You wouldn't," I said, as if to challenge Tony.

"Oh yes, I will!" he taunted back.

That's all I needed.

Gary started the interview, and after a couple of minutes I entered the shot with the cake and Gary transitioned to Helio's birthday.

"Helio, we understand that this is a very special day for you."

That's when he spotted me with the cake and played it out graciously and with the right amount of South American humility.

"Oh, no, I'm so speechless!" he said with a smile.

When I handed the cake to Tony instead of Helio, I saw Castroneves's eyes register that something was about to happen and it was definitely not what he expected.

As Kanaan inched closer and closer to Helio with the cake raised at his face, de Ferran joined in egging Tony on.

"Please! No!" begged Helio.

And then WHAM! Cake in the face!

Castroneves was covered in frosting and cake and tried to make nice with the joke. Helio's a pretty good jokester himself—but for a moment, after realizing that his buddies had turned the table on him, Helio was speechless. It became a piece of tape that was repeatedly aired throughout the 2003 IndyCar racing season and certainly will air many years to come as part of ABC's Indy 500 coverage.

But when the TV cameras were shut off, Helio went directly to me.

"You!" he said laughing and pointing his finger at me. "You got me! But Jack, I promise I will get you back someday!"

He hasn't yet, but Helio has a very long memory and I'm sure when I least expect it—WHAM!—Helio will get even.

❖❖❖

FINAL RETIREMENT

In 2003 when I first heard that Tony Kanaan was hurt in the Firestone Indy 300 in Japan and that Michael Andretti was going to have Mario shake the car down and test the car at Indianapolis, I was excited. You forget how old Mario actually is because he still looks like he could get behind the wheel. There was always a twinkle in his eyes that hinted at the belief, "You know what? If I wanted to do it, I bet I could still do it."

Mario Andretti, in Michael's racing suit, gets ready to test one of Michael's cars for Indy in 2003.
(PHOTO COURTESY OF THE INDIANAPOLIS MOTOR SPEEDWAY.)

On Mario Andretti's first day of practice, I turned on the radio and heard he was posting really fast laps.

"Yes, the old man has still got it!" I thought as I turned on ESPN. Then I saw him take flight. As I watched the car suspended in air, all I could think was: "Get out of the car, Mario, before you get killed."

When he got out of car after the crash, it was the first time I ever saw Mario look like he knew that his career was over. The twinkle was gone. I could see in his eyes that he knew things were over as he talked to the press from Gasoline Alley.

It wasn't fear. Fear had nothing to do with it because, in my opinion, Mario Andretti is more fearless than A.J. Foyt. I don't know why, but I could see that he came to grips with his mortality; I thought he had decided that this "Racing Grandfather" thing is kind of selfish and stupid. Somewhere in midair or when he landed, he came to the realization that he wanted to be there for his grandkids.

❖❖❖

SAY CHEESE

Helio Castroneves had a hilarious stunt pulled on him in 2003.

A bunch of guys set up the prank so that Helio would get pulled over while driving Gil de Ferran to the racetrack. The two of them got stopped right in front of a billboard congratulating Helio for his two Indy 500 victories.

The cop walked up to Helio and asked him for his license and registration. The officer pretended he didn't know Helio while Helio was being very polite and apologized for going too fast.

Then all of a sudden he realized he could get a ticket. The cop wasn't budging.

So he started to say he was running late and his boss was Roger Penske. He told the cop he was headed to the racetrack and he said humbly, "Hey, I am Helio Castroneves. You may have heard of me."

"Nope," the officer said, filling out the paperwork.

Now Helio realized he was getting a ticket, and he started offering some passes, or a helmet. You see, Helio had a lot of speeding tickets and he didn't want to get another one.

The cop looked at him and said, "Are you trying to bribe me, sir?"

Helio got really nervous; he thought he might lose his job with Penske over problems with the law. He began to panic.

The entire time, good old Gil de Ferran was just sitting in the car stone-faced. All of a sudden, the cop looked at Helio's passenger and said, "Hey, you're Gil de Ferran! You drive for Roger Penske!"

Now Helio lost it. He was just angry. The cop and de Ferran kept on prodding him right to the point where they finally let him know the jig was up. He had been caught on Candid Camera.

❖❖❖

A LAST GOODBYE

For what was said to be his last race at Indy, Michael Andretti started the month jovial as he went about his business.

Michael Andretti exits his race car after throttle linkage problems in 2003. (AP/WWP)

But May was still not without its issues for him in his new role as an IRL team owner. Tony Kanaan had a broken wrist, and Michael had a third car for rookie Dan Wheldon that needed attention.

And for Michael the driver, there were farewell press conferences and functions because everyone wanted to say goodbye to Michael Andretti.

He was dealing with all of it the best he could, but by the time race day rolled around, you just knew that Michael wanted it all to be over. And it was—quickly.

Michael's car developed throttle linkage problems and went to pit road for the final time on lap 94. There was a feeling of disappointment in the air even before Michael shut off his engine for the last time. He was still sitting in his car while others were speeding by out on the track.

A throng of reporters crushed into the tiny pit area reserved for Michael's car, and his PR staff tried to keep all of them in control to allow Michael a few moments alone with his crew before he faced the press.

I was right there with the crew and was a bit surprised that nothing special happened. Like so many Indys before, Michael got out of the car before the race was over.

He told his engineers what happened, and they told him what they discovered had happened to the throttle linkage that caused his retirement from the race. Then Michael took a cap with his sponsor's name on it and turned and faced the media for his farewell interviews.

Michael said goodbye with the same dignity that had marked his entire Indy career. There was no bitterness—maybe a bit of disappointment.

Then Michael went to his garage in Gasoline Alley, conferred with his family, changed into his street clothes, and came back out to root for the other cars on his team that were still in the race.

Michael was always a real champ.

❖❖❖

A DAY IN THE LIFE

Rick Rinaman has been at Indy even longer than I have. He's been part of Roger Penske's talented crew for almost 30 years, and although most guys his age (mid-50s) look to move off of the over-the-wall crew to a less physically demanding spot behind pit wall, Rick has stayed put.

He most recently commanded Gil de Ferran's team. It's a cliché to say guys like Rick are the ones who toil behind the scenes, but let me give you an example of what guys like Rick Rinaman dealt with on race day in 2003.

4:30 A.M.

Rick and the entire Penske crew meet in the lobby of their downtown hotel. In addition to it being race day, it is also checkout time. These guys have called the hotel home for more than three weeks. But tonight, win or lose, they will be headed back to their Reading, Pennsylvannia, shop to get ready for the next IRL race.

In Rick's case, there's a test scheduled for Monday and Tuesday. So he makes arrangements for other crew members to take a lot of his luggage back to Reading while he has stuffed some clean gear into a small bag that he will carry on Roger's jet that evening for when he and a couple of others head to the site of the test.

Checkout goes without a hitch, and four vans pull up under the hotel's portico and the team loads for the drive to the speedway.

Because their day is being documented by ABC-TV, they have a police escort. Normally, the caravan must make its way to the track on its own. But today is a special treat.

5:00 A.M.

Wheels up! Two motorcycle patrolmen pull away with the Penske Team in tow. Lights are flashing and sirens wailing.

But Rick is already going over the checklist of what he and his guys must do when they hit the speedway grounds. Rinaman gets distracted when some of the guys in his van start hooting.

One of the motorcycle patrolmen has stood up on the seat of his Harley and is doing a no-hands run down the street leading the procession.

5:25 A.M.

The Penske guys are not the first team to Gasoline Alley, but they quickly make their way to their complex of garages and start to work. Every move is detailed on a minute-by-minute checklist that Rinaman has received from Penske Racing president Tim Cindric.

Item by item, the crew double-checks lines, hoses, electrical fittings, wing angles, etc. As each item is checked, the team member who checks it initials the prep sheet.

7:15 A.M.

Gasoline Alley is starting to buzz. Race day fans lucky enough to have the correct credentials start to fill the aisles and the yellow-shirted security guards' whistles are blowing.

Inside the Penske garage, Rinaman spends some time with the engineers to make sure that all of the fuel mapping and computer programs have been loaded into the Toyota engine that will propel Gil de Ferran.

7:27 A.M.

The crew fires up de Ferran's car. They check the telemetry and make sure that there are no leaks or other unexpected concerns. Once the engine gets to temperature at 7:32 a.m., the crew shuts off the engine and sends the car to the IRL inspection line.

8:00 A.M.

The car has gone through inspection, and although it is ready to roll out onto pit road, it remains inside Penske's garage. It's a prop for a VIP garage tour that includes important sponsor representatives from Phillip Morris and the host of other corporate partners that Roger Penske entertains with his IndyCar efforts.

Rinaman plays his role like a seasoned pro. He answers a couple of questions, but all while he and a couple other crew members are positioned at the corners of the car to make sure that one of the VIPs doesn't accidentally hit a wing or knock something askew.

8:35 A.M.

The car and crew roll out onto pit road. Their spot is very near pit exit.

While Rick has been patrolling the garage, some of his team have already set up the pits and his tire expert has already laid out their Firestone Firehawks in sets ready for use in the race.

8:40 A.M.

Rinaman and his over-the-wall team practice pit stops. Four tires off, four tires on, simulate filling the car's tank with 35 gallons of methanol.

But Rick and his crew also practice pit miscues. They go through the scenario for if the car stalls: A crew member at the ready quickly inserts the outboard starter. They practice no tire changes.

It's like watching a football team practice—offense, defense, and special teams. Every possible scenario is run through and practiced.

All of this goes on while thousands of fans are filling the pits for a last look at the cars. In the middle of all this practice, Rick has to politely move some fans away from the car so that practice can continue.

It's all very polite and Penske-like.

"Excuse me, ma'am. We need to get to that wheel to practice our pit stop," he said.

"Oh, I'm so sorry. I didn't realize I was in your way."

"That's OK. We'll be done in a couple of minutes, and the car will be here for pictures."

9:10 A.M.

Picture time has arrived. One by one, fans get in a line to kneel next to Gil's car for a quick photo. Rinaman and company watch their baby with a protective eye.

Roger Penske stops by the pits to chat—just small talk.

10:00 A.M.

The cars are positioned on the front straightway of the track. Rinaman's car is situated on the inside of the fourth row, three cars to a row. De Ferran has completed his required appearances and prerace briefings, and just before he heads to the pagoda for driver introductions, he stops to chat with Rick and the crew. It's small talk. These guys have done this dance so many times that as they get closer to the 11:00 a.m. start, the more they move into a familiar race mode.

10:51 A.M.

Mari Hulman George gives the command to start engines. Rinaman gives a final thumbs-up to de Ferran and as the red and white liveried G-Force pulls away, Rick thinks about the countless hours the team spent trying to decide whether they should run the G-Force or the Dallara chassis.

He also flashes back to Phoenix, where Gil crashed with Michael Andretti and suffered back and neck injuries that kept him out of the cockpit when the team ran in Japan.

This is Gil's first race back, and Rick's experience reminds him to expect the unexpected.

11:00 A.M.

The 500 is underway!

Throughout the course of the race, Rinaman and his crew perform without a hitch.

❖❖❖

A TEARJERKER

When Gil de Ferran won Indy in 2003, it made it easy for him to retire because now he had done everything he wanted to do, everything he could be asked to do in a race car—short of going back to recapture what was lost in Formula One career aspirations.

When Gil pulled his car into Victory Lane, a quick look at him showed that he was spent. Crew members surrounded the car cheering and congratulating their driver, and Gil just sat in the car.

His firesuit was soaked in sweat; the tire soot and grime from 500 miles of racing covered him and the car.

When he finally was free of all the safety equipment, he eased himself out of the car and rested on the roll hoop.

As he pushed up from the cockpit, you could see that it was a painful process because of a recent back injury. Once up and visible, Gil let out a tremendous yell of jubilation and thrust his arms skyward.

I thought the winner's interview would be like many before—a happy driver recounting his day, thanking his crew and sponsors.

I asked Gil, "How tough was it to win this race?"

He teared up.

He started to talk about an injury he had suffered in Phoenix and the battle the team had had in picking a chassis to run at Indy. The tears just started flowing.

I was taught very early in my TV career that it wasn't important to ask a thought-provoking question. Instead, I was told it was important to evoke an answer that is.

When Gil started to cry, there was nothing more to say. It was the Indianapolis 500, and his win had brought this hero to tears.

❖❖❖

A DAY IN THE LIFE REPRISE

2:22 P.M.

Rick Rinaman returns to a spot he'd visited first with Rick Mears, then with Al Unser Jr., and then again with

Emerson Fittipaldi—Indy's Victory Lane. It's de Ferran who gets all of the attention, but right there is the crew and Gil takes a moment to publicly thank them for all of their hard work.

"You will never know just what a team effort today was," de Ferran said, nearly overcome with exhaustion.

Rinaman just smiles.

4:00 P.M.

Rick and the crew have changed out of their fire-retardant pit uniforms and back into their garage gear—pleated slacks and white short-sleeve shirts that say Marlboro Team Penske on the back.

They start to dismantle the garage that they have called home since early April.

All of the tools and equipment must be loaded into two 55-foot tractor-trailers. This includes some of the stuff that sets Penske's shop apart: diamond-plated Gladiator brand walls and storage bins, plasma televisions, plush office furniture, and dividers.

Each crewman has an assignment, and despite the fact they have won the biggest race of the year, they go about their business. Over in the corner is the winner's garland, along with some open bottles of champagne.

Milk was the beverage in Victory Lane, but when the crew returned to the garage, some well-wishers brought the bubbly for them to enjoy. No time for a long celebration, though.

6:45 P.M.

Rinaman is tying up the last loose ends as the crew packs up the garage. He is a few minutes away from head-

Gil de Ferran and crew pay homage to the yard of bricks after their 2003 win. (PHOTO COURTESY OF THE INDIANAPOLIS MOTOR SPEEDWAY.)

ing to the airport when Tim Cindric comes in and gives him a victory present—the next day off. Tim tells Rick the test is canceled. Rick will join Roger Penske, Tim Cindric, de Ferran, and others at the formal awards banquet on Monday.

❖❖❖

ANOTHER ANDRETTI?

Will there be a third generation of Andrettis at the Indianapolis Motor Speedway? From what I hear about the racing progress of Michael's son, Marco, I'd say absolutely.

Will Marco be rushed to Indy? No. Instead Michael will let Marco get there like he did—at his own pace. You

are going to see Marco Andretti arrive just like Michael did—ready to do battle.

For Italians, it's a dynamic of our heritage. We always want one generation to hand the legacy on to another. The proudest moment for a father is when the son decides to follow in his footsteps. But Italians want their sons to do better than they did. There's always that warm feeling when a son decides what you do is worthy and that it is his desired career.

That's the case with Michael. If you want to get a smile or get really deep thoughts from Mario Andretti, just talk to him about Michael. Ditto for Mike on the subject of Marco. When Mario analyzes Michael, you aren't sure if it's the pride of a father speaking or the respect for another race car driver.

Michael is going to be a great mentor, when, not if, Marco comes to Indy and there's another Andretti racing in the 500.

❖❖❖

THE ALL-AMERICAN

In 1998 Eddie Cheever won the Indianapolis 500. Then "America's Race" quickly morphed into Brazil's Race. After Kenny Brack, a Swede who took the 1999 honors, Brazilian drivers swept the top spot for four straight years. There was no reason to believe that things would be different in 2004. That was until Buddy Rice was hired by Rahal Letterman Racing to replace Brack, who came within inches of losing his life in a horrible crash at Texas at the end of the IRL's final 2003 event.

Rice is an All-American kid who grew up in the Phoenix area. Born (appropriately in July of the United States' Bicentennial—1976) "Little Bud" got his first taste of racing following his dad, "Big Bud" to Arizona drag strips where Big Bud competed weekly.

"It was a father-son deal," Rice recalled. "I really was more interested in baseball, but then I started karting with my dad and caught the bug."

As a high school senior Rice faced a crossroad.

"I was pretty good at baseball, good enough that some colleges were looking at me. That's when my dad told me I had to decide between baseball and racing."

Rice chose the latter and started his ascent that ended with his first Indianapolis 500 start in 2003 for fellow Phoenix resident Eddie Cheever. Rice was one of only 18 Americans in the 33-car starting field that year. The race had gone global, and fewer Americans were earning Indy car rides. Rice brought his cap turned backward, baggy shorts, and edgy T-shirt "X Gen" lifestyle to the big time and provided a refreshing breath of fresh air to a series that was quickly starting to look more like a Formula One race than what Indy always was—a slice of the Midwest. After

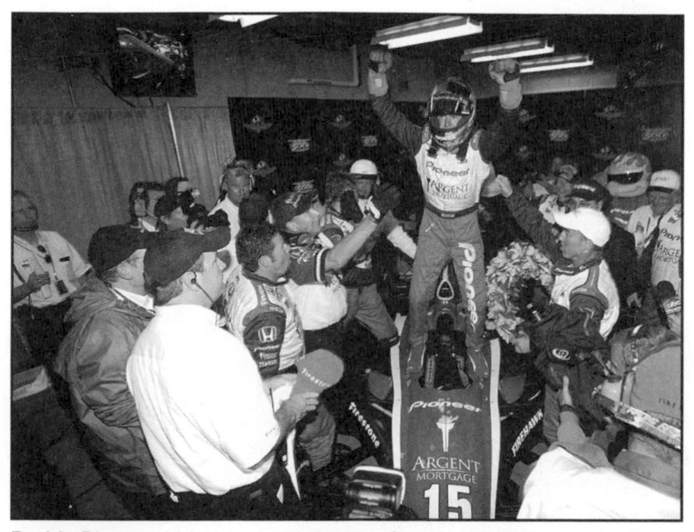

Buddy Rice and his team celebrate after he brings home the Borg Warner trophy.
(PHOTO COURTESY OF DAN HELRIGEL/INDIANAPOLIS MOTOR SPEEDWAY.)

finishing 11th at Indy and making an inauspicious run during the rest of the season, Cheever released Buddy from his contract citing his team's need for a leader and more aggressive driver.

When the 2004 season dawned, Rice was still unemployed. His offseason job search was unsuccessful until Rahal Letterman Racing called offering him a ride while Kenny Brack recovered from his accident at Texas Motor Speedway. Brack was still on the mend and Rice was in the seat of the Pioneer/Argent Mortgage car for Indy practice and qualifications. It didn't take long for Rice to show why he belonged at Indy. Rice grabbed the pole position of 222.024 mph beating out Dan Wheldon and Dario Franchitti for the front row. Some people were mildly surprised by his pole run. In fact the Monday following pole weekend during one of our TV production meetings, I suggested that we do an in-depth feature on Rice, his baseball background, and what looked to be Cinderella rise to the top spot.

"Rice will be 10th by the 10th lap," the producer chided. "Those Andretti Green cars will bury him once the race starts."

I held my tongue, and after the meeting I shared those comments with Rice, but he was not upset.

"Jack," he told me. "Let them keep thinking that. I don't care what they think. I know what I've got!"

When race day dawned, there was a good chance that rain would postpone the event for a day. A steady rain was falling a couple of hours before the race, but then at about the time the race should have started, the clouds broke. When the green flag finally did fly several hours later, it was obvious that this Indy 500 was likely to be an Indy 250. There was no way that the rain would stay away long enough for the full 200 laps. Once the field got beyond the halfway point, it was an official race, and the teams up and down pit road shifted their tactics.

Rice and Rahal didn't need to. Right from the start they raced every lap as if it were the last one. There was no pacing or conserving equipment.

Afterward, Buddy told me, "It had nothing to do with the rain. The car was just so good that I just drove it."

Rice led 68 of the first 97 laps until a red flag stopped the action. After yet another drying period, the race went back to green and Rice went back to work. The steel gray sky had darkened and black—and I mean *black*—clouds were creeping over the Speedway. It was just a matter of time before the race would end. Some teams tried pushing their pit envelope, gambling that the race would end just beyond the halfway mark. Others like Rice stayed on schedule and after a pit stop would just rush as quickly as possible back to the front. Around the 150-lap mark, Rice showed his real muscle.

Coming down the front stretch he darted to the inside of two cars screaming at 220 mph just inches off the inside concrete wall and made a three-wide pass on the way to the lead! He was so close to that wall that the crewmen on the other side assigned to signal their drivers with their signal boards dove for cover.

With just 28 laps left of the scheduled 200, Rice roared back into the lead passing Adrian Fernandez for the top spot. Two laps later the yellow flew again for rain, and four laps later it was over. A downpour brought out the red flag, and Rice delivered Rahal and TV talk show host David Letterman their first Indy 500 win as car owners.

The downpour forced winner's circle celebrations inside a Formula One garage adjacent to Indy's traditional winner's area. But that didn't dampen the celebration. Rice downed the quart of milk like it was nectar from the gods.

While the Goirdon Pipers bleated out tunes on their bagpipes and the Borg Warner trophy rested on Rice's rear wing, Rice, owners Bobby Rahal and Letterman, and general manager Scott Roembke celebrated their win. America had a new hero. Buddy Rice, unemployed racecar driver was now an Indy 500 champion!

❖❖❖

ADDED BONUS

Before the race, the CEO of one of Buddy Rice's sponsors Argent Mortgage offered him a bonus.

"He told me that if I won the race, he'd buy me any car I wanted," Rice told me.

The Argent CEO kept his word. Rice got the car of his choice. No foreign Lamborghini or Ferrari for this champion.

"I got a kick-ass 1949 Mercury hot rod. Lowered, chopped and channeled," Rice told me later that year. "They tricked it out with an awesome Pioneer sound system and I drive it everywhere when I'm home in Phoenix."

It was an All-American choice from an All-American guy.

❖❖❖

RAHAL'S REDEMPTION

When Bobby Rahal won his Indy 500, it came six days after the scheduled date and was overshadowed by the terminal illness of his owner, Jim Trueman, who died just days later. I always felt that Rahal missed out because so much attention was paid to his owner's battle to hang on long enough to fulfill his own dream. Now Rahal was back—as a car owner, and he took it all in wide-eyed wonderment.

❖❖❖

THE BOOTH

If there ever was a year when I wanted to be covering Indy's pit road, it was 2004. For 20 years that had been my beat. My home. A place where I could feed off the adrenaline generated by others and share in the Indy experience.

In 2004 I was reassigned to the broadcast booth as an analyst/storyteller. When ABC first informed me of their decision to "relocate" me, I admit that I was excited. Television has always developed their pool of analysts from former participants of the sport they were televising. The one recent exception is John Madden. For the most part though once their active careers were through, former stars could stay connected to the sport they loved through TV.

In IndyCar racing, you had to be a driver to get a shot upstairs. For 40 years, ABC's Indy 500 coverage included guys like Phil Hill, Jackie Stewart, Sam Posey, Bobby Unser, Danny Sullivan, Larry Rice, and Scott Goodyear.

Experts who spent their life in the sport like Chris Economacki were relegated to the pits. I always thought that limited the coverage. When Fox TV took over NASCAR, they did it right. In addition to Darrell Waltrip, they added Larry McReynolds to provide the pit perspective to their telecasts. So, when my call "upstairs" came, I welcomed the challenge.

Boy was that a big mistake. TV ratings for Indy car and the Indy 500 had plummeted in recent years so everyone had an opinion on how to improve them. That gave rise to individual agendas and some serious in-fighting between ABC and ESPN. I know most think that because both networks are owned by The Walt Disney Company that their relationship is harmonious but that's not the case.

ABC had a 40-year track record at the Indianapolis Motor Speedway. ESPN resented that and constantly battled for control of the property. For 2004, that meant the creation of the ESPN way and the ABC way.

I was part of the ABC way, and as an ABC guy the bull's-eye on my back was pretty big. Although ESPN provided their nightly coverage of the action at Indy and coverage of some of the qualifying, ABC had the crown jewel—the race.

The tension was heavy in the production compound. The backbiting was the worst I'd ever encountered. It was common knowledge that ESPN wanted to make wholesale changes as far as talent was concerned. Paul Page, who had anchored ABC's Indy coverage for almost 15 years, was not high on ESPN's list. They wanted a younger hipper approach and did little to disguise their desires.

This made me both uncomfortable and wary about my new role.

"Jack, just go upstairs and tell stories about the drivers and the crews," was the advice that ABC's coordinating producer Curt Gowdy gave me. "You don't have to analyze."

Unfortunately, it didn't take long for me to realize that good analysis was what was really needed. Paul and Scott Goodyear always saw things from only a driver's perspective. Neither had spent any significant time on pit road so they were weak when it came to strategy and tactics, two things that play a huge role in all racing but are never more important than racing in the Indianapolis 500.

"Drivers are all alike," Indy car engineer Jeff Britton once told me. "They get only part of the picture by design. With all the telemetry that we have on these cars, we put them on a need-to-know basis. A lot of what we do, they don't need to know."

Rain wiped out all of our qualifying programs, so race day dawned with Paul, Scott, and me waiting to finally get a chance to actually talk about race cars on the race track running at speed. But that was not to be.

Heavy rains delayed the start of the race and what was to have been a one-hour prerace presentation dragged on until it became a four-hour lead up to the eventual command to start engines. The race actually started about the time we were originally scheduled to go off the air and we still had 500 miles to call.

Four hours and 27 minutes later Buddy Rice pulled into a makeshift Victory Lane sheltered from the rain that cut the race short by 22 laps, and my marathon stint in the broadcast booth was over. Our telecast was record breaking. We were on the air eight hours and 27 minutes—the longest live sports telecast ever aired by ABC Sports with the exception of the award-winning Olympics coverage.

It didn't take long for me to realize that being upstairs was not going to work out. I saw things differently than my fellow boothmates. This was going to be Rice's race. His strategy was perfect. Others gambled on the weather. Some short-pitted an early stop so that once the rest of the field started to pit, their driver would move up in the standings more by default than by pure speed.

There were times when I felt like such an outsider that I just sat back and watched the race. Paul and Scott had a style, and I quickly discovered that it wasn't mine.

Being disconnected from the pits for the first time in my Indy career left me hungry for information. I couldn't just wander up to a crew chief and talk about strategy. I was landlocked—out of the weather and insulated from the very things that I always feasted upon while covering the 500.

Feeling less a sense of accomplishment than a sense of survival the three of us wandered down the steps from our perch exhausted.

"Get out!"

Whistles shrilled in the background as a Speedway security guard ushered us to an area beneath the concrete grandstands.

"There's a tornado heading our way," he barked.

A possible tornado was the perfect capper to a day that left me disappointed. I tried to do my job as instructed but couldn't escape the frustration that being locked upstairs and not down in the elements rubbing elbows with pit crews created all day.

Mike Pearl, ABC's headman, tried to raise my sprits when we finally were released from our tornado holding area and slogged back to the TV compound.

"You did exactly what we wanted you to do," he told me. "Congratulations. Good job!"

I knew I could contribute more. I also knew that ESPN would never agree to me being part of the booth team. That day, I left the track knowing in my heart that when 2005 rolled around, Page would be gone and I would be sent back to the pits. I wasn't an ESPN guy.

❖❖❖

THAT GIRL

I always thought that Sarah Fisher would be the first woman driver to win the Indianapolis 500. That was before Danica Patrick burst on to the scene.

Patrick enjoyed the landscape altered by the accomplishments of Janet Guthrie and Lynn St. James. As a teen, she tore up the karting world racing against all comers with a single-minded drive best exemplified by an interview she gave to ESPN when she was just 15.

By then she was an established champion. As she showed off all her first place trophies, Danica told the interviewer, "I really don't keep my second or third place trophies. Just my first place ones." She then ushered her guests to a spot in her room that was empty. "This is where I'm going to display my trophy for winning the Indianapolis 500."

Danica's dad, T.J., and her mom, Bev, have been with the 23-year-old right from the beginning. She definitely gets her drive from her father.

"She's never satisfied," T.J. told me. "If she won a race by a full lap, she'd want to know why she couldn't have won it by two. I wanted to know why she didn't win it by three laps."

The plan for Patrick's run in Indy cars was struck back in 1998 when she packed up all her belongings and shipped off to England to race in the British Formula Vauxhall series. England is a long way from tiny Roscoe, Illinois. But a drive to succeed eased the homesickness, and winning did the rest.

By 2001 Patrick was one of Europe's top female drivers and had a much desired American cache. But, there still was that empty space in her room back home.

Bobby Rahal signed Patrick to a developmental contract. Rahal put her in the Toyota Atlantics Series and started the Indy 500 grooming process. It was all calculated. Nothing was left to chance. While her training took place on the track, this pert brunette with striking girl-next-door looks built a rep off track.

She hosted a show on Spike TV and posed for photo layouts in several magazines. Patrick was hot and she worked it.

"I don't have anything to be ashamed of," she told me. "I am a girl, you know, and what most people don't realize is that I really like dressing up and acting like a girl."

When practice opened for the 2005 Indianapolis 500, that girl was already a cultural icon. She'd shown her muscle by leading 32 laps in the race preceding Indy, at Twin Ring Motegi in Japan. Mainstream media were tracking her progress. Not just the beat guys who were at every IRL race, but CNN, NBC, *Sports Illustrated*, *People* magazine. Patrick's potential brought coverage from all fronts.

Rahal knew this would happen. That's why the previous year, he sent Patrick to Indy to observe.

As Rice was winning the Indy 500, Patrick was there with his team (as well as Rahal's other entry for Vitor Miera) sitting in on every driver-engineer meeting, attending every public relations function, and most importantly absorbing everything that both elevates Indy and debilitates a rookie. As she sat through the rain and the down time while the crew worked on the car, she saw the never-ending demands that drivers juggled as they tried to qualify, prepare, and race in the biggest race in the world.

By the time 2005 rolled around, Patrick was both an Indy rookie and a seasoned veteran. The only thing she had not done was actually drive at Indy. That all changed on May 9. Patrick breezed through her rookie orientation. She was a full two mph faster than her nearest fellow rookie, Ryan Briscoe. Her best lap of 222.741 mph marked her as a pole contender. The 169 practice laps (the most of the 11 rookies who participated in R.O.P.) were geared toward winning the pole.

Throughout practices the week before Pole Day, Patrick was always at or near the top of the speed charts. Between laps, she dutifully met with the media and did the traditional Indy dance. Rahal's training paid off. Wedged between extended sit-down interviews for the swelling media horde, sponsor functions, and early plans for her fall nuptials, Patrick was tuning her Dallara for a shot at Indy's pole position.

"We want to take her along slowly," Patrick's chief engineer Ray Leto told me that week. "She wants to go quicker all the time, but she's learning that at Indy with all the practice time, you try things, then talk about what they did, and then make changes and go through the process again. She's an impatient driver sometimes."

"I don't blame her," T.J. Patrick snapped back to me when I told him about my conversation with Leto about his daughter's preparation efforts. "She's been groomed for this month. Sure, she's a rookie, but she's ready for this!"

When the day before pole position qualifications ended, Patrick looked ready. She clocked in with a final practice speed just a little more than one mph slower than Tomas Scheckter's fastest lap of the day. Her 226.769-mph speed made her a solid prospect for Indy's top spot. Now all she had to do was wait.

❖❖❖

THE RACE FOR THE POLE

Pole Day dawned at Indy awash in rain. Typical late spring weather put a damper on what many hoped would be a record-setting day. Despite the rain, fans showed up and many of them were there to watch Danica Patrick.

"She's the bomb," one rain-slickered fan told me as he splashed around behind the Speedway Pagoda. "That girl can drive!"

In Gasoline Alley, almost every door to garages was shut. Rain brings out groundhog-type behavior. Teams will crack open their doors, see the rain, and retreat back into their garages, sequestered from the reality that means they will have to wait to perform.

In Patrick's garage, her Rahal Letterman team was making busy work. Patrick ducked inside a couple of times but for the most part tried to stay away. There was nothing she could do about the weather. But it meant that her quest to make history was about to be put off for a day.

"Attention in the garage," the public address in Gasoline Alley boomed. "Today's activities have been canceled."

When I caught up with Ray Leto, Patrick's engineer and in many ways her coach, he just shrugged his shoulders.

"Another typical day at Indy," he said. "Guess we will just have to do it all tomorrow."

The postponement of pole qualifying meant that twice the number of spots (22) would be up for grabs the next day. Originally, Indy's qualifying was to be spread out over four days with 11 spots locked in each day and, for the first time, with teams allowed to make multiple attempts (up to three a day) to better their position.

"Will that take some of the pressure off you guys?" I asked Leto. "I mean you're certainly one of the 22 fastest here."

"Not us," he said. "Our goals are high, and that won't change."

The rainout did help the other members of the Rahal Team. Buddy Rice had backed into the wall earlier in the week and wasn't yet cleared to race by the Speedway medical staff. We would learn later that his crash injured some neck ligaments and the defending 500 champ would not race, but as the rain fell, Rice was still hopeful that he could go the next weekend.

At the end of the day, Patrick's PR staff released a statement about her reaction to the qualifying postponement.

"I'm not sure how we're going to be. I'm sure I'll have knots all over my body," she said. "You don't want to do anything that makes you doubt the car. You have to have the most confidence in the world in your car to keep your foot down, so I think the first lap has got to be a confidence builder. The second lap, you've got to push it a little bit and go after it after that."

Her words were prophetic.

Sunday morning I understood what Leto was talking about. In the final practice session before qualifications, Patrick led the pack with a speed of 229.880 mph, the quickest clocking of the month! Her nearest competition, Dario Franchitti, was almost one-half mph slower.

When Patrick put up that lap, the crowd cheered. Up and down pit road, team members from other teams mentally logged the speed and all of us knew that this year's qualifying was going to be history making.

Sam Hornish put up the time to beat early. The second car out, Hornish's four-lap average speed was 225.847

mph. But that speed fell to Tony Kanaan immediately. Kanaan stitched together a four-lap average of 227.566 mph.

Six cars later, Patrick took to the track, and the crowd cheered lustily. That cheer grew when Tom Carnegie announced that her warmup laps were quicker than Kanaan's.

"And she's on it," droned Carnegie as Patrick took the green flag for her official attempt.

She went into turn 1, and all I heard was the crowd collectively sigh, and I knew something was wrong even though I couldn't see the track. Her race car had washed out and she came close to crashing, just barely saving it from the wall. At that point I figured her qualifying run would be a washout with the bobble choking her aggression.

"Lap number one for Danica Patrick," Carnegie announced, "224.920 mph."

In fact, I was very surprised that she continued on. As she completed her second lap, Carnegie announced her speed. It was just four one-thousandths slower than Kanaan's second lap.

"Hey, wait a minute!" I said to myself, "she's got some moxie."

I'd seen many rookies take days to recover from a near crash, and here was one who took less than a lap.

Her third lap was two one-thousandths faster than Kanaan's third lap, and her final lap of 227.860 mph was a full three-quarters mph quicker than Kanaan's final qualifying lap.

When she came to a stop on pit road after her run, you knew that she was hot. Patrick hopped out of her car and with her helmet still on pleaded with Leto and Rahal.

"The car just stepped out on that first lap," she said. "I know we can win the pole. Let's get back in line."

As I listened to her plead, I hoped that Rahal would grant her wish. I pictured her topping Kanaan's time and imagined the buzz that would surround the race if she won the first spot. But that's when Rahal did what any great coach would do.

"We'll talk about it," he murmured to his driver.

But then he turned away from her and signaled to Scott Roembke, his team general manager, that Patrick's speed would stand.

"I feel like I had the pole in my hand, and it slipped out," she told me her voice tinged with disappointment. I was almost as disappointed. "But, you know, I think what happened in turn 1, almost losing it and hitting the wall and then catching it, might have just done me some good as far as earning respect."

In the end, Kanaan's speed stood up and he took the pole. Hornish and Scott Sharp took the other front row spots, and Patrick had to settle with fourth on the grid.

"I know I'm going to be facing a tough situation on race day," she said. "I have no doubts that I can do it. I usually race better than I qualify. Let's hope that that happens here."

❖❖❖

THE 2005 RACE

Right from the drop of the green flag, you just knew you were going to see history. Tony Kanaan took the early lead from his pole spot, but Danica Patrick stayed right in the hunt. She hovered in the top five throughout the first

50 laps and went through her first pit stop with the fanfare of a winner.

When Patrick pulled her car onto pit road for her first pit stop, her Rahal Letterman team went through their rounds with no miscues.

"Reset fuel," was the only radio traffic between Patrick and her engineer.

It looked like it was smooth sailing for Rahal.

❖❖❖

HER FIRST MISTAKE

Her second pit stop resulted in a rookie mistake when a normal day in the pit exploded into frenetic activity. As she shifted to get out of the pits, her car lurched forward and died.

"It won't go," she pleaded in frustration.

The crew stood there shocked for a few seconds before they began to move. Patrick froze as the men around her automatically responded to the problem as they had practiced countless times. I looked into the cockpit and all I saw were Patrick's eyes—they weren't scared; they were frantic.

All of a sudden the radio hailed, "We'll get it."

Leto's voice seemed to soothe his young driver's nerves because her eyes relaxed and she refocused on getting the car moving. Her eyes looked ahead at the track and she seemed determined to make this her Indy 500.

As quickly as it happened, her Honda fired up and she was back on the track.

❖❖❖

ERROR NUMBER TWO

Danica Patrick, slowly and methodically with the help of Ray Leto, her engineer, worked her way back to the head of the pack. But on lap 155, Patrick committed her second error.

I was jarred from what I was doing by the commentary of ABC's play-by-play man Todd Harris.

"Danica has crashed! The IRL's only female driver has spun! What a terrible turn of events."

It was a total surprise to me; it should not have happened because the field was under the yellow flag and the group was just about to restart. But the timing was Patrick's big break.

Her spin kept the yellow out and allowed her to duck into the pits for a quick fix. Once she got there, the crew replaced the nose on her car. In the safety of the pits and under the privacy of her helmet, Patrick let out all of her frustration. She let out a few screams, calmed herself down, and jumped back into the pack without losing a lap.

"She dodged another major bullet," I thought as she zipped by. "Maybe the stars are aligning for her and a date with Indy's Victory Lane."

❖❖❖

HER SHINING MOMENT

By lap 172 Danica Patrick's moment in history came—except most at the track didn't realize it at the time. The yellow flag came out on lap 170 and through pit stops Patrick became the first woman to lead the Indy 500! When she inherited the lead though, the caution (and pit stops) was still under way. The shuffle gave her the front spot, but fans didn't realize it until she took the field down for the green flag on lap 174.

The crowd roared as Patrick took the green and screamed into turn 1. They didn't stop for 12 laps until Dan Wheldon passed her on lap 186. Then one lap later the yellow came out again.

"Danica," intoned Ray Leto to his driver, "make this restart count."

Patrick and her team were desperately trying to save fuel.

"Don't be afraid to use the button," Leto added.

IndyCars have a button that increases the fuel mixture and horsepower to the engine called the overtake button. Drivers use it when trying to pass or on restarts to give them a momentary burst of speed. When Leto gave the okay to his driver to use the button, you just knew it was going to suck up valuable fuel and likely erase all of the fuel that they saved. But this was the Indy 500, and after all that she had survived, Leto wanted his driver to go for it.

"I grabbed third on the start and I passed him," Patrick said. And when she did, the entire crowd stood on its feet.

"Was I aware? I did notice a few people standing actually. I saw some arms waving. But I was very focused

Team co-owner Bobby Rahal congratulates his driver Danica Patrick, who became the first woman to lead a lap at Indianapolis 500. She crossed the finish line in fourth place, the best finish by a female driver.
(PHOTO COURTESY OF KAY NICHOLS/INDIANAPOLIS MOTOR SPEEDWAY.)

on my race," explained Patrick after the race. "I could see people standing and waving a little bit just in my peripheral vision because us drivers have that fine-tuned peripheral vision, but I didn't hear them."

I did! The roar was the loudest I'd heard at the Brickyard in many years. Usually that sort of affirmation is reserved for a winner on his final lap, but Patrick brought it out with 10 laps to go!

I also knew that Patrick would need some luck to become the first female winner of the Indianapolis 500. She had been saving fuel for so long, and Dan Wheldon was given the green light to use all of the fuel that he needed to overtake her and win his first 500.

With six laps to go, that's what he did. Wheldon streaked by Patrick and went on to win his first Indy 500.

"Saving fuel had to override everything else," Patrick said afterward. "I also was getting a little bit loose at the end. The car was starting to move around a little bit."

On lap 197 her teammate Vitor Miera got by and so did Bryan Herta.

"It was frustrating to be leading the race with so few laps to go and not be able to finish hard and just hang out up front and win the thing. But I also knew that I was not in the same strategy and something had to give."

When Wheldon drove across the yard of bricks on the final lap, Patrick was the fourth car behind him—the highest finish for a female in Indianapolis history.

❖❖❖

MICHAEL GETS HIS, THANKS TO LIONHEART

When Michael Andretti exited his race car in the 2003 Indianapolis 500, I knew that he was ready for his next challenge. I also knew that despite never winning the 500, he had assembled a potent team. Dan Wheldon, Dario Franchitti, Bryan Herta, and Tony Kanaan had a unique chemistry. Michael and his co-owners Kim Green and Kevin Savoree had a good package—Honda power and the Dallara chassis. Kanaan roared to the 2004 Indy car title, and they were big players in the 2004 Indy, falling short to Rahal Letterman's "super sub" Buddy Rice.

Indy in 2005 would be different. Wheldon went into the month as the point leader. After day one of qualifying was washed out, Wheldon trailed his teammates in qualifying, posting a disappointing 16th. His mentor and teammate Tony Kanaan took the pole, and you could tell that this bothered the Englishman.

Wheldon had been all but unstoppable in the early races, winning the opener at Homestead and then scoring convincing wins at St. Petersburg and Motegi, Japan. He opened 500 practice with the fast lap of the day at 226.808 mph, but when it counted, all he could muster was a four-lap average of 224.308 mph.

"I was so upset by my qualifying performance," Wheldon said. "The Indianapolis 500 is the most competitive race in the world, and I think it's probably going to be the most exciting race for a long, long time. That's what we're building for, but with the competition level it's going to be a tough one."

Maybe being denied a top starting spot was what Wheldon needed. On the top of his helmet is the royal crest of Richard the Lionheart, a medieval English king known for his bravery.

"I'm not sure if it's a compliment because he basically goes to battle with his heart rather than his head," Wheldon once told me. "It's something I added back in my karting days."

"Lionheart" was ready. From the drop of the green flag you knew that Wheldon was on the move. Although Kanaan led the field, Wheldon started his move to the front. By lap 60 he was in the top five and stayed there until fast work in the pits with a little less than 50 laps to go gave him a lead over Vitor Miera, Franchitti, and Kanaan.

When the field went back to green, "Lionheart" swapped the lead with Miera and Indy's female phenom Danica Patrick. This was when Wheldon showed his newfound maturity.

When he first came to Indy in 2003, he flipped his car while admittedly trying to block Sam Hornish with just a handful of laps left. After that race he got a well-earned butt-chewing from his owner Michael Andretti and admitted, "I drove like an idiot."

In 2004 he was no match for the destiny-driven Buddy Rice and settled for a third-place finish. This time would be different.

All afternoon Wheldon calmly went about his business. When Patrick passed him for the lead on lap 190, he didn't act like an idiot. Instead he coolly recognized that Patrick was low on fuel and could not run her car as hard as he could.

By lap 194, Patrick slowed to conserve fuel, and it was Wheldon and his Lionheart helmet that streaked by the crowd of 300,000 to lead the final laps and take his first Indy 500.

His joy was palpable. He banged on the steering wheel and you could see he was shouting in celebration from behind his helmet.

CHECK THAT OFF OF THE LIST

At the beginning of the season Dan Wheldon and his Andretti Green teammates sat down and set goals for the team. First was to win the series' first road course event (that their owner Andretti Green Racing promoted) in St. Petersburg. They did with Wheldon, Kanaan, Franchitti, and Herta sweeping the top four positions (which was their second goal) Their third goal was to win the Japan race—Wheldon did that for the second year in a row—

Co-owner Kim Green, Dan Wheldon, teammate Bryan Herta, and co-owner Michael Andretti take in the view from Victory Lane. This was Andretti's first taste of an Indy win.
(PHOTO COURTESY OF MICHAEL VOORHEES/INDIANAPOLIS MOTOR SPEEDWAY.)

and finally "win one for Michael"! Deliver Andretti his long awaited Indianapolis 500 victory.

That fourth goal was the top on Wheldon's individual list.

"I don't think people ever realized how important Indy was to me personally," Wheldon told me. "It was the only American race I watched as a kid. I'm sure that I'll win a lot of championships and races in my career, but none will hold up to winning Indy!"

Those comments helped me to understand why the young Englishman was overcome with emotion when he got to winner's circle. I waited on him and just let him get it out before doing the winner's interview. The emotions changed quickly from retrospection to jubilation when Andretti got to the ceremonies.

Andretti embraced his driver, and I thought I might see a repeat of his father's Indy-winning car owner Andy Granatelli's kiss. Instead they shared a private word. "Thanks!" was all Michael said.

❖❖❖

GOT MILK

While Dan Wheldon posed for pictures, I grabbed Michael Andretti and congratulated him.

"I thought you were going to kiss him," I said chuckling.

"I almost did!" Andretti confessed. "Man, am I glad this is over. No more Indy curse!"

With that I retrieved the remains of the traditional bottle of milk that was resting on the sidepod of Wheldon's car and offered it up to Michael.

"Here," I said. "You deserve this."

Michael looked me in the eye and took a giant gulp.

"I waited a long time for this," he said before lifting the bottle to his lips. "Thanks!"

❖❖❖

ALL I GOT WAS THIS T-SHIRT

Danica Patrick's efforts earned her the front cover of *Sports Illustrated* that week and far more media attention than the actual winner. By the time the Indy cars arrived at Texas the following weekend. Buddy Rice and Vitor Miera sported T-shirts that said "Danica's teammate" and "Danica's other teammate." Wheldon sported a T-shirt as well. It said, "Actually won the Indianapolis 500."

❖❖❖

THE PLOT THICKENS IN 2006

Just when you think you have Indy figured out, events remind you that Indy is as Eddie Cheever once said "a living breathing organism."

Michael Andretti earned his first-ever win after years of frustration and disappointment when Dan Wheldon took Andretti Green into the 2005 winner's circle. Wheldon also captured the season-long championship—and then announced that he was leaving the team.

Wheldon's departure set in motion one of the biggest offseason stories ever, a story that has two parts. First was the decision by Andretti to return to Indy as an owner and a driver.

"I think I'll have as good a shot as I've ever had to win the Indianapolis 500," was the way Andretti announced that he would drive in the 2006 Indianapolis 500.

I couldn't agree more. His team has developed into the top team at Indy, and Andretti's time away from driving gives him new perspective. He'll have nothing to prove. Instead, he can just race and in the process share an experience few fathers get to enjoy, which is the second part of the story.

I said earlier that I expected another Andretti to make his mark at the Brickyard. The race in 2006 will be the one. Andretti's son Marco will fill Wheldon's spot on Andretti Green.

Marco Andretti spent 2005 racing in the Menards Infiniti Pro Series, the IRL's feeder series. In six races he went to winner's circle three times. His first run in IRL came after the 2005 Michigan race.

"He went out and did a fantastic job for the team," Andretti recalled, obviously proud of his son's performance. "He was right on pace right away, had great feedback. The team was impressed with him."

We will just have to wait and see if another Andretti face joins Mario's on the Borg Warner trophy. I wouldn't bet against it.

❖❖❖

2007 INDY 500

May 2007, was a month of domestic turmoil for me. A divorce was hanging over me and as much as I had hoped that my time at the 500 would provide me with a degree of escape, it turned out to be a period of reflection that put my self esteem to the test.

Few knew about my personal issues. But as the month progressed more and more of my friends knew something was awry. One of them was Dario Franchitti.

Franchitti joined the IRL from CART in 2003 and became a cornerstone to Michael Andretti's AGR IndyCar operation. He was quiet and stayed to himself.

"Jackie? Are you OK?" Franchitti asked me one day as we waited on pit road for practice to resume. To this day I cannot tell you why, but I shared my situation with the Scotsman. "I don't know what I'm going to do," I told Dario. "I thought spending time here would alleviate the internal turmoil, but instead it's made me fearful."

"What?" asked Dario.

"I don't know what I would do if I ever lost this job. It's become my life and now I am paying the ultimate price again."

"Look, Jack," Dario said firmly. "You belong here. You are part of this place and you are not going anywhere. Before I got here, I knew you were part of the Indy 500 and that will never change."

At that moment, my fears and uncertainty vanished. I felt like Sally Fields, when she won her Oscar and blurted out "You like me! You really like me" to the Academy members in her acceptance speech.

Then, as Dario stood to get into his car he grabbed me and said, "I love ya, man."

I walked away with a lighter heart and ready to take on all my challenges.

When the opening cannon for the 91st running of the 500 went off at 6:00 a.m., an overnight rain left a damp weight to the air and cloud cover promised that Mother Nature was not finished with her watering.

Days like that drive strategists and engineers crazy. There are a host of mechanical adjustments crews can use to compensate for the weather, but how to race (i.e. Should we pit out of sequence? Should we short pit? etc.) are always gut calls dictated by unfolding conditions.

That was the ace, played by Dario's strategist, the late John Anderson. This savvy vet knew that it was highly unlikely that the race would go the full 200 laps. When the green waved, he started Franchitti out on a conservative drive that played to his driver's wheel house.

"Jim Clark is my hero," Dario once told me. "He was disciplined and always explored every opportunity that a car presented to see which ones he could exploit."

While Helio Castroneves, Tony Kanaan, Marco Andretti and Scott Dixon all took turns leading, Franchitti went off sequence in his pit stops and prepared for his weather gamble.

"We will see how it works out for us" was Anderson's answer to my question about why he went out of sequence. Listening to Franchitti's radio transmissions you sensed a questioning tone to his transmissions. But like his hero, Franchitti followed Anderson's orders to the "T" and inherited the lead by lap 74. But his time to pit out of sequence was near and by the time rain brought out the red flag on lap 113, Dario was trailing and his teammates filled the top three spots (Kanaan—Marco Andretti—Danica Patrick).

The red flag went on for hours. Kanaan waited out the red wondering if the race would be declared official and give him his first 500 win.

But, down in the Franchitti pit, John Anderson stuck to his strategy and waited to see if he would be able to play it out completely.

At 6:00 p.m. the race resumed. Now no one was certain whether or not the race would go the full distance. Most team raced like it would.

But Anderson, kept Franchitti on his off sequence strategy. They waited patiently.

"Stay out! Stay out!" was Anderson's transmission to Dario when the caution flew on lap 151 for Marty Roth.

Franchitti complied and as the leaders ducked onto pit road for fuel and tires, Dario inherited the lead followed by Scott Dixon (who also stayed out for track position).

The race went back to green five laps later but not for long. Kanaan and Jaques Lazier tangled bringing out the caution.

Now the sky was a steel gray. You knew that time was running out on this race and when the race resumed on lap 163, it was Franchitti that brought the field down for the green. Behind him though, was chaos. Marco Andretti tangled with Dan Wheldon, and the young Andretti found himself flipped over and sliding down the backstretch.

The yellow came out yet again and as the heavens opened up to the heaviest downpour of the day, Dario gingerly guide his car through the puddling water to take the Checkered Flag.

In Victory Lane, Franchitti sipped the milk, expounded on his remarkable drive and waved to the crowd. John Anderson was absent. He remained at Dario's pit and passed on the celebration. I asked him later why he did not go to Victory Lane. He told me his job was done. "I knew what we had accomplished and that was good enough for me," he said. "Besides," he continued, "I was tired. It was past my bedtime!"

As I was finishing up the ABC TV interview with Dario in Victory Lane, my month at Indy came full circle. "That's it from Victory lane," I said on the air preparing to throw back to the ABC broadcast booth. "Dario, congratulations!" That's when Dario winked at me and said, "I love ya, man!"

❖❖❖

2008 INDY 500

In the spring of 2008, a dispute that tore the IndyCar Nation apart came to an end. CART, from which the Indy Racing League split away from in 1996 was disbanded through bankruptcy just before they were to kick off their season. CART's remaining assets were purchased by Tony George and the Indy Racing League which for CART purists was sacrilegious. Acrimony ran high among CART protagonists.

When the Indianapolis Motor Speedway opened for their annual May activities, CART's top drivers were housed in Gasoline Alley right next to the stars of the IRL.

Oscar winning actor Paul Newman put his signature to a letter distributed to CART fans asking then to support what was now called unification. He urged them to come to the Indianapolis 500.

As practice unfolded, it was apparent that the IRL regulars had the upper hand. The one exception was Will Power, who was driving for KV Racing. Power won the very last CART race, the Long Beach Grand Prix run the same day as Danica's win in Japan. His CART team, Walker Racing was one of the unification casualties and Power was picked up by KV racing for the IRL season.

"These cars are just so different," Power confided to me that May. There were other issues too; while Power and the other CART drivers publicly supported unification, the rides that they picked up for the 2008 season were in essence spare cars previously owned by IRL Teams.

Pundits have always criticized the IRL Dallara chassis as a "spec car" with little innovation. But that Indy showed that there was a lot of work to get one of those spec cars to keep up with the Penskes and the Ganassis.

Dario Franchitti had gone off to drive Chip Ganassi's NASCAR Sprint Cup entry which meant that he was not at the Speedway to defend his win of the year before. His departure from Andretti Green Racing was seen by some as an effort by Ganassi to disrupt the momentum that Team AGR had developed with Dario in the AGR fold.

I never put any credence to it, but the end result was an Indy 500 for Chip Ganassi's Team Target operation.

Scott Dixon won the race from his pole spot and unification, while underway, was not complete. No CART driver finished in the Top 10 that day. They would spend the rest of the season catching up with their IRL brethren.

❖ ❖ ❖

2009 INDY 500

The family of Helio Castroneves believes that "God has a plan." That belief was severely tested when Helio and his sister were indicted for alleged tax evasion. A lengthy investigation involving Helio and his first Penske contract saw Helio and his sister starting the 2009 season in a federal courtroom in Miami instead of on the IndyCar circuit.

When Helio signed his first contract with Roger Penske in 2000 it was days after Penske's first choice, Greg Moore, died in a crash at the California Speedway. Helio was headed back to Brazil after being told that his ride in CART with Carl Hogan Racing was ceasing operation.

Penske wanted to get the deal done ASAP. So, existing contracts that had been drawn up for Greg Moore were altered and Helio's name was inserted. Some offshore payments raised the IRSs' ire and Helio found himself trying to escape deportation as well as jail time.

"Those were the hardest days of my life," Helio told me. "I was confused and scared. I always tried to do the right thing. I relied on others because I was not an American and now I was facing the possibility that I would never be able to do what it was I loved so much ever again."

Then there was his sister Kati, "She was pulled into this and that hurt me so much. All she ever did, from day one was love and support me. Now with a small child and a family of her own, she was on trial too."

The 2009 season started while Helio sat in that courtroom. Will Power was brought in to replace Helio until he could return to the cockpit.

On April 17, a federal jury acquitted Helio and Kati. When Helio exited that courtroom he said, "It's been a long seven weeks. I'm a foreign person and I've been judged in a foreign country. I'm very thankful to have received a fair trial. I do love this country."

By the time the Indianapolis 500 rolled around, the euphoria of Helio's acquittal had yet to subside. Now he was part of a three car effort at the 500—a thank you to Will Power for his efforts while Helio was sidelined.

Fans and friends could not stop themselves from wishing Helio well. But those close to him sensed something was different. When I asked him, Helio still smiled but he said, "Jack, you don't go what I have gone through without it having an effect. I learned that God has a plan for all of us and He is always in control."

I knew that Helio practiced his Roman Catholic faith. Father Phil DeRea, the IndyCar priest, would often say a special Mass for Helio if his schedule kept him away from the one open to all members of the racing community on race day. But, I didn't think much about Helio's walk with his Maker until he said those words to me.

If I had described pole day that year instead of witnessing it, you would accuse me of "going Hollywood."

When the call came for Helio to put in his ten mile, four lap qualifying run, the weather was not optimum. His speed was good enough for the front row. But it was on the outside of the front row instead of the inside. As the day progressed, another Penske/Ganassi battle ensued. Dario Franchitti sped his way onto the front row but by mid-afternoon, the conditions were improving and you knew that something was going to give.

With the diminished number of cars entered in the race over the past few years, more and more of the top rung teams showed little fear in withdrawing their time and trying again.

As long as there was no weather threat, it made for an interesting strategy.

But when Helio formally withdrew his time to make a second pole attempt, some wondered if it was a good risk.

We got an answer real quick.

Helio posted a 224.864 mph average—better than his teammate Ryan Briscoe and Ganassi stalwart Dario Franchitti.

When the gun sounded to end Pole Day, Helio jumped into the arms of Tim Cindric and held on tight. For a moment he looked like the "Little Helino" that would often do the same to his dad when he was victorious in a go-kart race.

Race Day was just another chapter in the Helio Castroneves saga.

He led 66 of the 200 laps and beat Dan Wheldon by 1.9 seconds and Danica Patrick to win his third Indianapolis 500, only the ninth driver to win three times.

Tears flowed everywhere. Competing teams signaled their congratulations. But it was the quiet moment in Victory Lane that defines Helio.

Before our interview with ABC, the winner's wreath and the traditional gulp of milk, Helio's mother Sandra bent into the cockpit and tearfully hugged her son. God indeed had a plan.

❖❖❖

2011 INDY 500 QUALIFYING

It was the Centennial edition of the "Greatest Spectacle in Racing" and from qualifying through the race itself, the 2011 Indy 500 was filled with memorable moments.

In qualifying it was the struggle of Andretti Autosport. The five car team of Marco Andretti, Danica Patrick, John Andretti, Mike Conway, and Ryan Hunter-Reay went into May with high expectations.

But it was immediately evident that something was awry with the team. Patrick's was the only car showing speed. The other four struggled through the limited practice due to rainy conditions and when qualifying weekend dawned, AA was clearly on the ropes.

All the Andretti cars were shut out of day one qualifying spots.

And, as Bump Day progressed, deep concern developed about the possibility that Patrick would miss her first Indy 500 in 10 years, along with Marco Andretti. These two were used to being in the upper rung in 500 qualifying.

At the end of pole day, Patrick actually made peace with the possibility that she might miss the show. "I thought to myself that if I didn't qualify that as terrible as that would be, life would continue," she told me.

One by one, the AA team took their shots at the 33 car field. John Andretti made it in, ditto for Danica. Conway, Andretti, and Hunter-Reay remained outside the field of 33 with less than an hour remaining in qualifying.

Conway had already won a race in 2011. His emotional victory at Long Beach earlier in the season brought his year full circle. In 2010, the Brit was severely injured on the last lap of the 500 when he crashed into Hunter-Reay on the last lap of the Indianapolis 500 as Hunter-Reay ran his fuel tank dry. Months of rehab and countless operations later, Conway came into the '11 Indy 500 fifth in points.

Unfortunately for Conway, the speed was not there. He used all three of his allotted qualifying attempts and failed to make the show. Hunter-Reay coaxed enough speed to qualify last on the grid with just one car left to go before the track closed at 6 p.m.. It was his teammate Marco Andretti. In a bizarre twist of fate, Marco bumped his teammate from the field leaving Hunter-Reay and his sponsor DHL out of the biggest IZOD IndyCar event.

Meanwhile, at the top of the charts was Alex Tagliani and his Bowers & Wilkins team.

Tags financed the team in 2010 with help from Rob Edwards and Allan McDonald, whom Andretti fired from his engineering staff the year before.

In 2010, Tags flirted with the top spot at Indy during qualifying. He was strong enough to move to the Fast Nine segment on Pole Day 2010 but fell short when the gun sounded.

In 2011, Tagliani was simply happy to have a car for the 500. Just weeks before the start of the season, his operation was on the edge of closing. Then, Sam Schmidt stepped in and purchased the assets and preserved the team.

From day one in practice, Tagliani found himself hovering near the top of every practice session. While most pundits focused on other teams and their preparation for pole day, Tagiani and his operation just kept tweaking their Dallara.

"We built upon what we learned in 2010," Rob Edwards told me. "The difference this time was that we were able to pool our set ups with those of Sam's other teams."

Tagliani easily qualified for the Fast Nine segment in 2011.

"I was not going to play it safe," recalled Tagliani. "We went into the "Fast Nine" with one goal; and that was to win the pole!"

One by one, the others in the fast nine took a shot at Tagliani. But when it was all over, the native of Quebec sat atop the scoring pylon with the pole slot for the 100th anniversary Indianapolis 500.

While Ryan Hunter-Reay contemplated his future as a non-qualifier for the biggest race in history, Alex Tagliani celebrated what he called his greatest racing accomplishment. "I can't believe it," gushed Tagliani. "It just goes to show you what hard work and a dream can get you."

❖❖❖

RACE DAY 2011

It was touted as the "Most Important Race in History" and the 2011 Indianapolis 500 did not disappoint.

From a pre-race that included the return of more than 250 living veterans of previous Indy 500s to the debut of double file restarts, the advance attention to the 2011 Indy 500 left you knowing that this one was going to be special.

When the green dropped, an early joust for the front spot went on between Alex Tagliani, Dario Franchitti, and Scott Dixon.

Conspicuously absent from the front were any of the Penske cars. Their May had been a disappointing one and race day proved to be no different.

When the first yellow came out, more than 300,000 fans waited in anticipation for the first ever side by side restart.

This new restart procedure had been the subject of much debate including a rumored boycott by the drivers. In the week leading up to the race, an informal polling of drivers failed to identify any who were in favor of the new procedure. "It's a recipe for disaster," groused defending race champ Dario Franchitti. "This is a one-groove race track and it's going to be ugly."

An 11th hour compromise saw side by side restarts in place but with the green flag coming in the short chute between turns 3 and 4.

So, when the green flew after the first caution, all attention was lasered in on the restart.

Not only did the restart go well, but it also shuffled the running order. Double file restarts were now officially a part of Indy 500 tradition.

The other pre-race story was Andretti Autosports decision to purchase the seat of Bruno Junqueira on A.J. Foyt's team. Ryan Hunter-Reay's qualification failure prompted Michael Andretti to buy out Junqueira and Hunter-Reay took over the seat in the Foyt 41.

The coyote red of Foyt's team car took on the canary yellow of Ryan's primary sponsor DHL.

All week, fans and pundits criticized the move. "It was not my decision," said Ryan. "I feel bad for Bruno. It wasn't the way I wanted to be in the Indy 500." But Ryan was in and Bruno returned to Brazil to watch the race he qualified for on TV.

Hunter-Reay was a non-factor in the race. After starting 33rd he never broke into the top 10 in the running order and he eventually finished 23rd, three laps off the winning pace.

By the halfway mark in the race, it became apparent that fuel conservation was going to play a big role in the race's outcome.

One by one, teams went off sequence and played different strategies.

After leading a big portion of the race, Franchitti went into fuel conservation mobbed dropping his lap speed by 3 mph. Tagliani crashed on lap 147 bringing the curtain down on his Cinderella story. Danica Patrick led at the mark as her crew put her into fuel conservation in hopes that the winning strategy they employed in Japan for her first IZOD IndyCar win would work at Indy.

But, as laps wound down, one by one drivers were force to pit road with empty fuel tanks. With less than 10 laps to go, rookie J.R. Hildebrand powered by Bertrand Baguette to lead.

Hildebrand replaced Dan Wheldon in the Panther Racing National Guard entry in 2011 after Wheldon was forced to sue for his back earnings. His season leading into the 500 was unspectacular but now, with laps running down, it was his opportunity to become the first rookie to win the 500 since Graham Hill turned the trick more than 40 years earlier.

Hildebrand was low on fuel. As he took the lead his crew started a dialogue with their driver. "Looks like we are going to have to pit," they told him with three laps to go. No sooner had Hildebrand received that transmission than he was told, "No! Don't pit. Save fuel."

The indecision from J.R's strategist had to be disconcerting at best. But it also filled the radio channel; a channel that also included Hildebrand's spotters.

"I'd passed on the high side all race long," recalled Hildebrand.

And as J.R. rode into the final turn and could see the checkered flag waving about a mile away, he chose the high side to pass back-marker Charlie Kimball.

Then, in a surrealistic turn, Hildebrand pushed out to the wall and crashed!

Before the caution light came on (which was about four seconds) Dan Wheldon ducked under his old car and streaked down the Indy front stretch to lead the race for the first time and in doing so win his second Indianapolis 500.

Hildebrand kept his foot in the throttle and willed his wrecked car down the front stretch bounded by the outside retaining wall to finish second.

Indy was Dan Wheldon's only ride for the 2011 season. After failing to land a full time ride in the series, he reunited with his old Andretti Green Racing teammate Bryan Herta for an Indy only effort.

Now both were the winners of the "Most Important Race in History" and in a most unusual way!

At season's end, Wheldon made another start in the season finale at Las Vegas Motor Speedway. He was part of a special promotion that put five million dollars on the line ($2.5 million for him and a matching $2.5 million to a lucky fan) if Dan started last in the race and won.

The brainchild of IndyCar's CEO Randy Bernard, it was hoped that the challenge would shine a spotlight on the Series finale.

It did. But in a tragic way.

On lap 11, Wheldon was collected in a horrific, fiery crash with 14 other cars. His car went airborne and crashed into the catch fencing in turns 1 and 2, his head smashing into a steel post.

After being airlifted to the hospital, the IndyCar nation learned that Wheldon succumbed to what was described as unsurvivable injuries. There is no way to describe the void left by Dan's death at age 33. Four Months before, "Lion Heart" won the Indy 500 and now he was gone; and our hearts wept.

❖❖❖

GASOLINE ALLEY

Recently, I was given a tour of Dreyer & Rinebold Racing's race shop expansion. As part of it, they have recreated the "old Gasoline Alley" replete with the white siding and green accents (I never could figure out why green—thought it was bad luck in racing!).

When I remarked to Robbie Buhl how neat I thought the touch was he said, "You know a couple of fans were in here last week and didn't know that Gasoline Alley once looked like this."

When I thought about it, I realized how much had changed within those sacred confines. When I first got to Indy in 1969, I was fortunate enough to score a "Silver Badge." Those things were gold! They gained you entrance into that place where everyone labored: Indy's Gasoline Alley.

The rows of garages were narrow and reminded me of what most hometown mechanics had in their backyard from which they twisted on their "Saturday Night Specials."

The one thing that separated these garages from the ones in a backyard was THE BLACKBOARD. Attached above each garage was a sign that told all who wandered through there which team occupied the space. Some were pretty fancy. Others? Well you knew that these were the teams who just hoped to make the race and pay for their May stay at the Brickyard.

All the signs shared one common denominator: they had a space where the team could chalk in their qualifying time. It was their proclamation that they were one of the elite 33.

There were no garage doors in the old Gasoline Alley. Instead, each garage's entrance was protected by a pair of large swing out doors that you'd see on an old barn. I always thought it was because this was an "Alley" instead of a "Garage" area.

When those doors swung open, it was magic. Crews would roll out their car and hook it to a garden tractor!

Each team that got a garage was given a tractor by the manufacturer. Eventually they would be replaced by more efficient towing units (golf carts and purpose built motorized rigs) but in '69 tractors still ruled.

The interiors where the teams worked for the month of May could be described as spartan. No real amenities except for tool boxes and parts. Almost every garage had a water cooler, a coffee maker and a refrigerator. Outside of that, An Econo Lodge looked like the Waldorf!

For a teen like me who dreamed about Indianapolis for as long as I could remember, simply wandering through this "Mecca" was better than a trip to Disneyland.

In 1981, I got my first chance to be part of the gang that lived inside those garages.

I was working for the Pepsi-Cola Company and worked a deal with Dan Gurney to sponsor his Chevy stock block Eagle that was to be driven by Mike Mosley.

Wayne Leary was the top wrench on the "Pepsi Challenger." He worked next to John Ward who was Gurney's engineer. They were a remarkable pair.

Ward was the thinker and Leary was the doer.

"Hey Pepsi Man!" was Leary's daily greeting, "We need more soda!" I was carting cases upon cases of Pepsi into this tiny garage and each morning, it was gone. It became a real mystery to me and one that I vowed to solve.

This team was special. It wasn't big and it didn't command the attention that others in Gasoline Alley did. There weren't many folks hanging around our barn doors. Instead, we went in and out unobstructed.

Each day of practice resulted in more wrench turning. Ward would sequester himself in a far corner of the cramped space and occasionally he would huddle with Dan Gurney. Then, Ward would issue Leary's marching orders. More changes to the car and another stab at some practice laps.

The biggest thing that I remember about Mike Mosley was how patient he was during this process. Other drivers held court while their car was worked on. Not Mike. If he wasn't sitting in a folding chair he was sitting on a bicycle getting sun just outside the garage. They call Scott Dixon "The "Iceman" but Mike Mosley was the original "Iceman."

By the time Pole day rolled around, the anticipation was at an incredible high inside that garage. No one thought we had a chance at the pole but Gurney told us all that getting in on Day One was an attainable goal.

While I took in all the festivities that kicked off Pole Day, Leary accosted me, "Hey Pepsi Man! We need more soda!"

Still stupefied by the quantity of Pepsi products that were being consumed by this randy group of 12 to 14 people, I made arrangements for another delivery of soda.

Under threatening skies, pole qualifying started four hours late. We were deep in the qualifying line and watched as A. J. Foyt logged a run of 196.078 mph. Based upon practice speeds, we knew he would not stay in P1, but the rain prevailed and eventually Pole Day was postponed to the next day.

Sunday was a total washout. Waiting in the rain in the old Gasoline Alley was quite the experience. A lot of crew guys waited things out in the Speedway cafeteria that was just outside the Alley and under the pit grandstands. For our team, it was some card playing and a lot of story swapping.

That rainy Sunday finally provided me the answer to my Pepsi mystery. I had worked up the courage to ask Leary about all the Pepsi we were consuming. "Come with me," said Leary; and we darted through the rain to the garage of Jim Hurtubise. "Hey Herc! shouted Leary, "Pepsi Man here wants to know where all his soda has gone!"

Hurtubise was one of those Indy legends that I grew up admiring. His budding career was sidelined in a fiery accident that left him critically burned. When doctors gave him the option of having refashioned hands that would allow him to engage in everyday tasks or hands that though claw-like would be able to grip a steering wheel, Hurtubise opted for the latter.

In 1981, Hurtubise was at Indy still trying to campaign a long outdated roadster. That year, he didn't turn any laps and I always wondered when I'd see that roadster that he called the Mallard take to the track. Inside his garage with Leary I discovered why Herc hadn't seen any track time.

"Well Pepsi Man," Hurtubise said laughing, "Me and Ole Leary here made a deal." With that he opened the hood of the Mallard revealing a cooler where the engine should be. "He brings me soda for this here race car and I give him beer!" Mixed between Miller cans were a lot of those Pepsi cans I was trying to locate!

"Them cases of Pepsi are currency!" laughed Leary as he doubled over, "Need Loctite—trade some Pepsi for it.

Need a sweatshirt? Hand over the Pepsi! Hell, boy we just drop off a few cases over at the watering hole and eat and drink for free!"

I was glad to know that I'd done my part to keep the Pepsi Challenger team well stocked and well fed.

Pole qualifying resumed the following Saturday. It didn't take long for Foyt's 196 mph run to fade. Bobby Unser took over the pole position with a four lap average of 200.546 mph.

When our car's turn came, Mosley motored out and started his 10 mile effort. His first lap was no where near Unser's, but if everything stayed together, we all knew that we would be solidly in the field.

Dan Gurney greeted Mosley in the photo area. "Great job, Mike. That will work." Mike squeezed out a four lap average of 197.268 mph.

As the original pole qualifying order took their turn on the track one by one, they failed to match that 197 mph posting.

At 2:00 p.m., the PA made the following announcement: Quaifying for the pole is over!

I looked up at the scoring pylon and no one had knocked the #48 off of the number two spot.

Mosley and the Pepsi Challenger had put a stock block powered car in the middle of the front row for the 65th running of the Indianapolis 500!

It was quite a celebration inside that our tiny Gasoline Alley Garage. "Hey Pepsi Man!" shouted Leary, "Fuck that soda! Here." He tossed me a beer.

The euphoria of that day evaporated the next time we took to the track.

It was Carburetion Day. A timed shakedown session that puts all 33 qualified (and the two alternates) out on the track for a final time. We didn't turn a lot of laps. We only had one engine and didn't want to put too many more miles on it. The only 500 miles that really counted would come on Sunday.

When the car returned to the garage, Leary, Ward, Gurney, and a couple of other crew members started pulling off the body panels that covered the engine area.

One of the cylinder heads came off and Leary peered into the opened engne. When he pulled out a pencil-thin flashlight and fished around an ominous cloud descended. Without a word, the twin barn doors were closed and the car was sheilded from prying eyes.

"I think we can get all the pieces out," whispered Leary to Gurney. "Well, do it," answered Dan.

Leary said, "I'll take care of it after everyone leaves."

Later that night, Wayne Leary replaced a push rod in that Chevy engine. He fished for all the miniscule pieces that might have scored the engine and buttoned up the car.

Years later, Leary explained, "We didn't have a spare engine and I knew that the engine would not last." "What could we do? The president of Pepsi was coming to the race with a plane load of bigwigs and Dan needed sponsorship money to race after Indy."

When all the Pepsi brass arrived on race day, they strutted around like they had hit the Lotto. "So what do you think Jack?" asked Joe Block, Pepsi's Vice President. "Wouldn't it be something if we won the Indianpolis 500?"

I remember thinking about how absurd that question was. I knew we would have no shot. But Mr. Block had no way of knowing that. He was standing inside a garage in Gasoline Alley but he really wasn't "INSIDE" the garage!

I took the Pepsi brass down into Turn One once the command to start engines was given. I wanted to get them as far away from our pits as possible.

"This is a great place to watch the early laps," I said.

"That's great" said one of them. "But, can we get back to the pit in time to watch a pit stop?"

Mosley was easy to spot on the track. The bright Yellow with red accents on the car made it stand out. With 400,000 fans watching and cheering, the race got underway and these first time visitors to the "Greatest Spectatcle In Racing" cheered their car on.

Sixteen laps later, Mike Mosley climbed out of the car and retired from the race.

He was credited with 33rd spot.

"Something got into the radiator," Leary explained to the Pepsi folks. As he said it, I caught his eye. It shouted "Don't!" So I didn't. I knew by keeping the secret I had gone from "Pepsi Man" to "Crewman."

The official box score for the Indy 500 listed the reason out as radiator failure. That will be the reason forever.

But, some of us know better.

Oh, the next week, Mosley took the same car (with a new engine) and started in the back of the grid at the Rex Mays 150 on the Milwaukee Mile. At the end of the race, the Pepsi Challenger was in Victory Circle! Mosley won and the brass at Pepsi got their trophy.

❖❖❖